Safety and Security at Sports Grounds

Steve Frosdick and Jim Chalmers

Published by Paragon Publishing

First published 2005

© Steve Frosdick and Jim Chalmers 2005

The rights of Steve Frosdick and Jim Chalmers to be identified as the authors of this work have been asserted by them in accordance with the Copyright, Designs and Patents Act of 1988.

All rights reserved; no part of this publication may be reproduced, stored in a retrieval system, or transmitted in any form or by any means, electronic, mechanical, photocopying, recording or otherwise without the prior written of the publishers or a licence permitting copying in the UK issued by the Copyright Licensing Agency Ltd, 90 Tottenham Court Road, London W1P 9HE.

ISBN 1-899820-14-0

Also available in paperback, ISBN 1-899820-16-7

Book design, layout and production management by Into Print (www.intoprint.net)

Printed and bound in UK and USA by Lightning Source

Dedicated to my grandchildren,

Kate, Mark and Sean,

for all the joy and happiness

you bring into my life.

Grandad Jim

CONTENTS

Foreword 7
About the Authors 8
Acknowledgements 10
Introduction by Jim Chalmers 11

PART I – STRATEGIC RISK 14
1.1 – Sports and Safety: Risk and Responsibility 15
1.1 – Commentary – Accountability for Public Safety in Sports Grounds 19
1.2 – Venues of Extremes: the PAF Approach 22
1.3 – Venues of Extremes: the Inner Zones 28
1.4 – Danger, Disruption, Finance and Fun 33
1.2-1.4 – Commentary – A Strategic Approach to Risk Assessment 37
1.5 – Constructing in Tight Corners 40
1.5 – Commentary – Stadium Redevelopment at West Ham United 43

PART II – SAFETY MANAGEMENT 46
2.1 – Safe Cracking 48
2.1 – Commentary – The Development of Football Safety Management 53
2.2 – Watching the Crowd 55
2.2 – Commentary – CCTV Requirements Specification 59
2.3 – Playing it Safe? 61
2.3 – Commentary – Police CS Spray and Crowd Safety 64
2.4 – Stadium Safety: Design, Management and Understanding Crowds 65
2.4 – Commentary – Case Studies in Stadium Design and Safety 73

PART III – SECURITY 74
3.1 – Violent Behaviour 76
3.1 – Commentary – Review of Spectator Violence 82
3.2 – Drink or Dry? 84
3.2 – Commentary – Alcohol in Stadia 89
3.3 – Royal Dutch: Integrated Ticket System 91

3.3 – Commentary – The Dutch 'COTASS' Project 96
3.4 – Unthinkable 97
3.5 – Security Pendulum 101
3.4 and 3.5 – Commentary – Venue Security After 11 September 2001 105

PART IV – SEGREGATION 108
4.1 – Keep off the Grass 109
4.2 – On the Fence: Balancing the Risks 114
4.1 and 4.2 – Commentary – Segregation from the Field of Play 118
4.3 – A Tale of Two Cities: Part I 122
4.4 – A Tale of Two Cities: Part II 127
4.3 and 4.4 – Commentary – Crowd Segregation at High-Risk Matches 131
4.5 – Problems on the Pitch 134
4.5 – Commentary – Player Misbehaviour Affecting Crowd Safety 138

PART V – STEWARDING 140
5.1 – Guidelines for Stewarding 142
5.1 – Commentary – Stewarding Guidelines 145
5.2 – Stewards Training Goes Multi-Media 147
5.3 – Dealing With Racism: Training Stewards 151
5.2 and 5.3 – Commentary – Stewards' Training 157
5.4 – Less Police: More Stewards 162
5.4 – Commentary – Higher Profile Stewarding 169
5.5 – Danger: Men at Work 171
5.5 – Commentary – Accidental Death of a Steward 175

PART VI – SEATING 177
6.1 – Ewood Effect 179
6.2 – Being There! 184
6.1 and 6.2 – Commentary – Improving the Atmosphere in All-Seated Stadia . 189
6.3 – Standing Up Again? 191
6.4 – Standing Debate 195
6.3 and 6.4 – Commentary – Persistent Standing in All-Seated Grounds 202
6.5 – Ten Years of Seating 206
6.5 – Commentary – the Ten Year Anniversary of All-Seated Stands 208

PART VII – EDUCATION 210

7.1 – Certificate of Higher Education in Stadium and Arena Safety 211
7.1 – Commentary – Job-Related Education 213
7.2 – Completely Safe.. 215
7.2 – Commentary – Evidencing the Competence of the Safety Officer 218
7.3 – Safety Techniques Shared 221
7.3 – Commentary – Lecture Visits to the Université de Technologie Troyes ... 223

Appendix 1 - Management Resources 224
Index 225

FOREWORD

In addition to his academic teaching and writing, Steve Frosdick has been writing articles on safety and security at sports grounds in professional magazines such as ours since 1994. The articles have been written in a journalistic rather than academic style and so have proved to be very accessible both to industry professionals and to the wider reader. Demands for copies from students and others led Steve to the conclusion that it would be useful if those articles were all brought together in one place, thereby providing the primary motivation for this publication.

The gathering together of material provided Steve with the opportunity to work with Jim Chalmers, whom he asked to review and update the articles. Jim has thus written commentaries to critique the articles, bringing them up to date as necessary.

This collaboration between a leading academic/consultant and a leading practitioner has produced some interesting results. There are areas where Steve and Jim agree with each other, but there are issues on which they clearly hold different views. It is unusual and rather refreshing to find co-authors airing their varying opinions in a public debate. As with many aspects of life, there are of course no 'right answers' to be found here. Rather there is discussion and debate, which is brought to life in the pages of this book.

The book has a primary focus on British football grounds, but the discussions will be of wider interest to an international audience. We are sure that anyone who reads our magazines would benefit from adding this book to their collection of professional publications.

We hope that students, practitioners and other readers will find the articles and commentaries useful, thought-provoking and interesting. Above all, we join with Steve Frosdick and Jim Chalmers in hoping that the book will help in safeguarding safety and security at sports grounds for many years to come.

Katie L McIntyre	Mark Webb
Editor	Editor
Panstadia International	*Stadium & Arena Management*

ABOUT THE AUTHORS

Steve Frosdick has been Director of IWI Associates Ltd since 1996. He has an MSc in Strategic Risk Management and is a Member of the International Institute of Risk and Safety Management. He is a Founder Member of the Football Safety Officers' Association and has held affiliations with several universities and consultancies.

Since January 2000 he has held a part-time academic post in the Institute of Criminal Justice Studies at the University of Portsmouth, where he teaches courses on safety and security at sports grounds and pursues his interests in the broad field of risk. Through IWI Associates, Steve has completed a wide variety of consultancy, research and training projects. His strategic risk management experience includes safety at sports grounds, events management, policing, programme and project management, strategic management, community and race relations, information security and general business risk. Clients have included the football authorities, various police organisations, conference organisers, universities and publishers.

From 1979 to 1995, Steve worked as a police officer in a wide variety of operational and support posts. Operationally, he served as patrol officer, custody officer, shift manager and operations manager. In other roles he gained experience in corporate strategy, project management, the management of risk, information management, quality improvement, performance measurement and cognitive assessment. He retired following a career break during which he successfully developed a consultancy and academic career.

Steve has an international reputation as an expert in stadium and arena safety and security. He has lectured and given conference presentations in the UK, USA, France, Portugal, Switzerland, Italy and Mauritius and has published over 50 articles, book chapters and other papers. He was the principal author and co-editor (with Lynne Walley) of the 1997 textbook 'Sport and Safety Management' (Butterworth-Heinemann) and is co-author (with Peter Marsh) of the 2005 textbook 'Football Hooliganism' (Willan). He is a frequent contributor to 'Stadium and Arena Management' magazine and has spoken at six of the annual 'Stadium and Arena' conferences since 1998. He is also the editor of the UK football authorities' 'Training Package for Stewarding at Football Grounds'. Although now a Southampton season ticket holder, he is a lifelong fan of Brentford.

Jim Chalmers served in the Birmingham City and West Midlands police services between 1959 and 2001, retiring in the rank of Chief Superintendent. Most of his police career was spent in uniform operational duties and at the time of his retirement he was the police commander at Aston Villa FC. He also had the unique distinction of being the police commander responsible for the Birmingham Super Prix, when the streets of Birmingham City Centre were turned into a motor racing circuit.

In 1991 he was appointed an inspector with the Football Licensing Authority (FLA) and served there until his retirement in 2003. As an FLA Inspector he was at the forefront of driving home the requirements for football stadia in the top two divisions to become all-seated in the aftermath of the Hillsborough stadium disaster. He was also closely involved in the design and planning of new stadium structures and related safety systems, particularly stadium control rooms and closed circuit television (CCTV) installations.

Between 1992 and 2003 he was member of the football authorities safety management focus group. He made a significant contribution to the development of safety management and stewarding practices at football grounds. He contributed to the football authorities guidance documents on these subjects and to the multi media stewards training packages developed by Steve Frosdick. He was also the author, co-author and contributor to various FLA guidance documents. These included spectator safety policies, contingency planning, exercise planning, briefing and debriefing, safety certification and the fourth edition of the Guide to Safety at Sports Grounds.

Jim has been a speaker on safety management and stewarding at the Institute of Local Government Studies at the University of Birmingham, at the police major sporting events courses and at safety officer training courses for the football, cricket, rugby and horse racing authorities. He has also spoken to government, sporting and police authorities in France, Holland, Germany, Cyprus and at the Council of Europe in Strasbourg. His expertise was recognised in 1996 when he was appointed to assist the government of Guatemala to make recommendations on how stadium safety could be improved after 88 football fans were crushed to death in their national stadium.

As a mature student he has obtained both a Certificate of Higher Education in Stadium and Arena Safety and a BSc degree in Risk and Security Management from the University of Portsmouth. He also holds a Level 4 National Vocational Qualification (NVQ) in Spectator Control.

He is presently the Deputy Safety Officer with Kidderminster Harriers FC. In 2005, his lifelong achievements were recognised by his peers when he was unanimously elected as President of the Football Safety Officers' Association.

ACKNOWLEDGEMENTS

Steve Frosdick wishes to thank the representatives of the football authorities and the many members of the Football Safety Officers' Association who have supported his work over the years. They have not been named to avoid compromising their anonymity. Steve would also like to thank those persons with whom he collaborated in writing the original articles reproduced in this book: Mel Highmore, Bob Rankin, John Sidney, Bruce Smith and Alison Vaughan.

Jim Chalmers wishes to thank his colleagues in the Football Licensing Authority; safety officers in football, cricket, rugby and horseracing; safety and security officials of the Football Association and Football League; and numerous members of football Safety Advisory Groups; for so freely and willingly sharing with him their knowledge and experience of safety and security at sports grounds over the past fourteen years. My special thanks however go to my co-author who for the past six years has been my tutor and mentor and without whom my contribution to this book would never have been possible.

Jim Chalmers and Steve Frosdick both wish to thank Mark Webb at Paragon Publishing for kindly agreeing to take on the production, illustration and marketing work for the book. We are also grateful to Mark in his role as editor of *Stadium and Arena Management* magazine and to Katie McIntyre of *Panstadia International* magazine. Thank you for having published the bulk of Steve Frosdick's work over the last ten years and for agreeing to write the Foreword to this book.

Acknowledgements are also due to the publishers of *Football Management*, *Football Decision* and *Police Review* magazines, which originally carried some of the articles included in this book.

Finally, we are both thankful for our wives, Alison and Pauline, who have patiently put up with our absences whilst we were away at matches or 'just popping into the office for a few minutes' (again) to study and write. Thanks also to Alison for copy editing the final draft of the book.

Steve Frosdick and Jim Chalmers

May 2005

INTRODUCTION
BY JIM CHALMERS

Anyone interested in the subject of safety and security at sports grounds – whether as a safety practitioner, sports regulator, police officer or academic – will know that over the years the subject has generally been hidden amongst inquiries into football stadium disasters or football hooliganism and violence. The first three editions of the Guide to Safety at Sports Grounds (the 'Green Guide') were written in response to specific disasters at football grounds. It was not until 1997 that the fourth edition was written without the pressure or lessons to be learned from a disaster.

It was in this same year that Steve Frosdick and Lynne Walley wrote and edited the book 'Sport and Safety Management'. This was the first academic publication to examine the full complexity involved in safely managing crowds at sports grounds. The book was not restricted to football and contained the thoughts of academics, sports regulators, safety practitioners and the police. It pulled together the many different disciplines involved in the safe management of a sports ground.

In the ten years between 1994 and 2004, Steve Frosdick has written many articles on various aspects of safety and security at sports grounds. Thirty-two of those articles have been brought together for the first time and republished in one place in this book. Jim Chalmers has reviewed each of the articles and written accompanying commentaries, making critical or reflective comments and bringing the articles up to date where appropriate. The articles provide a comprehensive record of the most salient issues which have had to be addressed, or which have tested safety management and security systems and procedures over the past decade.

This book will therefore interest anyone involved in the management or operation of a sports ground, particularly in a safety or security role. It will also interest those involved in the regulation of sports grounds, the police, academics and students whose studies bring them into the realms of sports grounds safety and security.

The book is divided into seven main parts. Each part has its own introduction, containing a short summary of the articles and commentaries within it. Each article is reprinted with its original citation and by-line shown and is followed by the relevant commentary.

Part I – Strategic Risk – comprises five articles dealing with the strategic management of hazards and risks at stadia and arenas, described by Frosdick as 'Public Assembly Facilities' (PAFs). The first article examines how concepts of risk, accountability and responsibility have developed in the aftermath of stadium disasters. The next three articles give a strategic overview of the diverse perceptions of hazards and risks and how these can be managed in the context of PAFs operations. Finally a case study is presented explaining how risk was managed at a large football stadium undertaking a major redevelopment programme.

Part II – Safety Management – comprises four articles relating to the safe management of a sports ground. The articles look at how safety management has developed over the years, examining various risks and aids to the management of spectators such as the use of Closed Circuit Television (CCTV) and the police use of CS spray in a sports ground. Finally, Part II suggests how the engineering risk management technique of 'Hazard and Operability Study' (HAZOPS) could be used to integrate design and management in a sports ground.

Part III – Security – comprises five articles about the security of sports grounds. John de Quidt (Chief Executive of the Football Licensing Authority) said that one of the fundamental lessons to be learned from the 1989 Hillsborough Stadium Disaster was that the needs of safety and security must not get out of balance. Part III examine the history of spectator violence and how the control of alcohol can impact on spectator behaviour in sports grounds. There is a detailed explanation of a club-orientated ticketing and marketing system (COTASS) introduced in Holland and a comparison of both the UK and American responses to stadium security in the aftermath of 11 September 2001.

Part IV – Segregation – comprising five articles, deals with issues relating to segregation from both the physical and management perspectives. The articles examine various methods of segregation and in particular the use of fences as a means of spectator control. A case study, examines present day segregation methods at football matches, and the final article in Part IV examines player behaviour and its effect on the crowd.

Part V – Stewarding – comprises five articles which cover the stewarding of sports grounds, particularly the football stadium. The first three articles outline early research into stewarding guidelines and explain how the training of stewards has been developed. The next article debates the evolution of high profile stewarding supported by lower profile policing. Part V concludes with a case study which examines the death of a steward at a football ground and the legal consequences for the Club.

Part VI – Seating – comprises five articles relating to the introduction of all-seated stadia in the aftermath of the 1989 Hillsborough Stadium disaster. The first two articles examine the lack of atmosphere in an all-seated stadium. The next two articles consider the reintroduction of standing terraces and look at the problems caused by fans who persistently stand in seated areas of the ground. Part VI concludes with a reflection on the tenth anniversary of all-seated stadia.

Part VII – Education – comprises three articles dealing with the qualifications and competencies of sports grounds safety officers. Particular mention is made of the Certificate of Higher Education in Stadium and Arena Safety, pioneered in 1999 by the University of Portsmouth. Part VII concludes with a description of the UK involvement in sharing sports grounds safety experiences with students at the Université de Technologie Troyes, France.

These above summaries perhaps explain why this book is so valuable to anyone interested in safety and security at sports grounds. No one book could ever deal with all aspects of these subjects, nevertheless Frosdick's articles provide a useful historical perspective and analysis of the major sports grounds safety and security concerns which have been arisen in the past decade. Chalmers has brought the articles up to date in 2005 with a critical review and practical commentary.

PART I – STRATEGIC RISK

Introduction

Part I comprises five articles dealing with the strategic management of hazards and risks at stadia and arenas, described by Frosdick as 'Public Assembly Facilities' (PAFs). The first article examines how concepts of risk, accountability and responsibility have developed in the aftermath of stadium disasters. The next three articles give a strategic overview of the diverse perceptions of hazards and risks and how these can be managed in the context of PAFs operations. Finally a case study is presented explaining how risk was managed at a large football stadium undertaking a major redevelopment programme.

1.1 – Accountability for Public Safety in Sports Grounds

Writing in 1996, Frosdick examines the implications of liability and responsibility for stadium and spectator safety given the growth of the 'blame culture'. There is a discussion of the principles of risk management and how consultancy can assist venue management in this process. Comment is made that whilst much of the article remains valid now, there has been considerable subsequent progress in the development of sports grounds safety management.

1.2 to 1.4 – A Strategic Approach to Risk Assessment

This series of three articles written in 1996 and 1997 present Frosdick's core ideas on the strategic management of hazards and risks in PAFs. He sets out the conflicting risk perceptions of four groups of stakeholders and demonstrates how dividing the PAF into five zones can assist in carrying out the hazard and risk assessment processes. The third article then illustrates how management can implement the various processes described. Comment is made that whilst the zonal approach has much to commend it, the management plan presents an excessively bureaucratic solution to the problem.

1.5 – Stadium Redevelopment at West Ham United

Frosdick examines the risks associated with major stadium redevelopment from the perception of a stadium safety officer. He uses West Ham United in 2001 as a case study, explaining the safety issues involved in converting a sporting venue into a construction site and then back again for the day of the event. The case study offers practical solutions to the problems this presents. Comment is made that the case study is a realistic depiction of the problems faced by any venue management in any redevelopment process. Examples are given of other redevelopments where spectator safety could have been prejudiced.

1.1 – SPORTS AND SAFETY: RISK AND RESPONSIBILITY

The original citation for this article is: Frosdick, S. (1996) 'Risk and Responsibility', *Panstadia International Quarterly Report*, Volume 3 Number 4, September 1996, pp. 34-36.

The disasters of the 1980s highlighted the importance of stadium safety. Steve Frosdick outlines the trends and how Staffordshire University is helping.

As a result of the disasters at Bradford football ground, the Heysel stadium and Hillsborough, the emphasis on safety in the British sports and leisure industry has sharpened in recent years. A large number of different government agencies and other bodies now offer detailed guidance on managing public safety in a wide variety of venues. The range of advice and recommendations is complex and clearly illustrates the growing importance now placed on the management of public safety risks.

The safety environment has been shaped by two key trends in philosophy and management. In the first place, the growing social awareness of liability and accountability issues has brought responsibility for safety into sharper focus. Secondly, the changes in policy for policing public events have seen the emergence of specialist safety managers, with stewards replacing the police in many venues. This article will explore these trends in more detail and outline how the Centre for Public Services Management and Research (CPSMR) at Staffordshire University has responded to them.

The Importance of Risk Management

Whilst total safety can never be guaranteed, good management means reducing the risks as far as is reasonably practicable. Risk is defined as 'the chance of exposure to the adverse consequences of future events'. There are therefore three elements to be assessed in a risk: the future event or hazard which may occur; the chance of the hazard occurring; and the adverse consequences if it does occur.

Managing risk is a two-part process. Risk assessment involves identifying the hazards, estimating their probability and consequences and evaluating their acceptability. The analysis will almost certainly reveal risks which have not been acceptably reduced, and this is where risk management is required. Appropriate action plans need to be prepared and resources allocated to progress them, controlling the risks either by reducing them, avoiding them altogether or perhaps transferring them to another party, for example an insurance company. Finally, ongoing monitoring is put in place to ensure the effective implementation of the plans.

But who is responsible for all this? The Home Office Guide to Safety at Sports Grounds – commonly referred to as 'the Green Guide' is quite clear that 'the responsibility for the safety of spectators at the ground lies at all times with the ground management'. Many sports facilities are multi-purpose, hosting pop concerts and the like as well as sporting events. And the Home Office and Health and Safety Commission Guide to Health, Safety and Welfare at Pop Concerts and Similar Events – 'the Purple Guide' – tell us that 'anyone who is directly responsible for the undertaking ... will have responsibilities for the health and safety of third parties affected by it, including the audience.' Finally, the draft Crowd Safety Guidance soon to be published by the Health and Safety Executive includes a statement that, 'ensuring crowd safety is a basic responsibility of venue managers, owners and operators'.

Let us therefore be clear that the responsibility for risk management and safety starts at the top, with the Chairman and Board of Directors, or their public sector equivalents, of the company or body running the facility.

Why does this matter? Well there is now a view that, because the world is a more individualist place, and because individual people feel a greater need to be protected from the effects of the world, the concept of risk has moved on from probability and consequences into the idea of risk as accountability, or risk as blame and liability, even without fault.

In the Fall 1990 Edition of *Daedalus*, the Journal of the American Academy of Arts and Sciences, George Priest wrote that, 'the more precise statement of the first principle of civil liability today is that a court will hold a party to an injury liable if that party could have taken some action to reduce the risk of the injury at a cost less than the benefit from risk reduction'.

The newspapers carry almost daily stories of large payments being awarded in civil damages for injuries sustained. And since many such cases appear to be settled out of Court, there is a suspicion that ability to pay is as important an issue as negligence in any pre-trial discussions about liability.

But it is not now just a matter for the civil law. In December 1994, British legal history was made. An outdoor activities company and its managing director were convicted of manslaughter following the deaths of four teenagers during a canoeing trip in Dorset. The managing director was jailed for three years, later reduced on appeal to two years. But he went to jail. And many more companies and individuals could find themselves facing criminal prosecution if the Law Commission's proposals for a new offence of corporate killing are adopted.

And the media's love of scapegoating means there is no doubt that, if public safety arrangements fail disastrously in the future, there will be a clamour for the owners and operators concerned to be called to criminal account. The best public liability insurance will not help the Chairman 'gripping the rails' at the Old Bailey.

Owners and operators therefore need to be sure that their safety arrangements are capable of withstanding the closest public scrutiny. They need to know how to assess and manage risk. They need to know how to document the process so they can produce the evidence to show their insurance company, a Civil Court, the police, a Coroner's inquest or even the jury in a criminal trial, that they did everything that they could reasonably be expected to do.

At the same time, they don't want to over-react with measures which are disproportionately expensive or which damage their customers' legitimate enjoyment. Nor are they required to do so.

The general principle is that risk should be reduced to a level which is as low as is reasonably practicable. Broadly speaking, those hazards judged to be of lowest probability and consequences will be accepted, whilst those which have been estimated as highest probability and consequences will be deemed to be intolerable and require action irrespective of cost. Where the boundaries fall between these two positions will usually be a question of management judgement, and will determine which hazards require action where something can be done at a cost less than the benefit of the risk reduction.

Changes in Policy

Concerns about the extent of their own liability, coupled with burgeoning demands for their services and increasing pressure on the public purse, have resulted in the police service seeking to reduce their commitments at public events, to concentrate on their public order and emergency management functions.

The Home Office Review of Police Core and Ancillary Tasks earlier this year proposed that, 'stewards could make a greater contribution to the policing of public events (such as football matches and pop concerts) and be more adequately trained for the purpose'. Reinforcing the trends in accountability, the Review also proposed that, 'organisers could assume fuller responsibility for the safety and behaviour of spectators and participants'.

At the same time, new legislation arising from European Community directives has resulted in general health and safety issues assuming greater prominence in public perception.

As a result of these policy shifts, venue owners and operators have increasingly had to buy in the necessary expertise, either by employing safety managers or by seeking guidance from consultants. They have had to move towards their own higher profile stewarding schemes, either by internal training and development or else by contracting out to a private stewarding company.

The Safety in Sports and Leisure Programme

This changing public safety and accountability environment was not being addressed within academe. Other universities working in the safety/order field were focused on different areas (for example, football hooliganism at Leicester, industrial health and safety at Loughborough). Having identified the gap in the academic market, the CPSMR at Staffordshire University established the Safety in Sports and Leisure Programme to provide a comprehensive package of research, consultancy and management development services.

To launch the programme, the CPSMR arranged a seminar which was attended by representatives of the governing bodies of various sports, regulatory agencies, football, rugby and cricket clubs. The dissemination of safety research findings is clearly important for a university programme and we have been pleased to give papers at four conferences this year, as well as to publish several articles and to participate in a panel discussion at the Soccerex '96. Most importantly, we have been commissioned by Butterworth-Heinemann to prepare an edited book on Sport and Safety Management, which is due for publication in March 1997.

Consultancy bridges the gap between theory and practice, and the CPSMR offers independent and confidential risk assessment, safety audit and inspection services to all sectors of the sports and leisure industry. But our main activity has been working with the Football Authorities and Football Safety Officers' Association to develop a multi-media training package for stewarding at football grounds. The package was published in June and is being taken into use throughout England and Wales from the 1996/97 season.

Over the coming years, we hope to see the programme continue to grow, supporting the CPSMR mission to help provide solutions to real management problems.

1.1 – COMMENTARY BY JIM CHALMERS – ACCOUNTABILITY FOR PUBLIC SAFETY IN SPORTS GROUNDS

In this 1996 article, Frosdick discusses how consultancy services could assist in providing solutions to safety management problems. At the time of writing the article, Frosdick was a consultant with Staffordshire University, where he led their Safety in Sports and Leisure Programme. It could therefore be argued that the article was a 'sales promotion', but to dismiss it as such would be a mistake, since the issues it raises are as relevant in 2005 as they were in 1996. The article was perhaps a premonition of growing concerns in the UK about the culture of 'blame and claim'.

With legislation on health and safety at work and on safety of sports grounds in place since the mid 1970s, why was it felt necessary in 1996 to remind senior sports grounds managers about risks, roles, responsibilities and accountability? When the 'Guide to Safety at Sports Grounds' (the 'Green Guide') was first published in 1973 it was made clear that responsibility for spectator safety lay at all times with ground management. That principle has been repeated in all subsequent editions. Why was it then that a police Chief Superintendent and Superintendent were charged with manslaughter following the deaths of 96 fans in the 1989 Hillsborough stadium disaster – rather than the ground management who had the responsibility for safety of the venue? It is not for me to comment on the legal or moral complexities of the case, but it should serve as a constant reminder that there is no room for complacency with regard to roles, responsibilities and accountability for spectator safety. For most public events, the police require a 'statement of intent' in which they and the event organisers will agree, in writing, their respective roles and responsibilities. It would now be a very foolish police commander who ever signed up to being responsible for spectator safety since this always has been and still remains one of the prime responsibilities of the ground management.

Frosdick reminds us of the process of risk management, but he overlooks the importance of the spectator safety policy, which should be drawn up by the ground management or event organiser. If no such policy exists, this should bring into question the management attitude towards spectator safety – and indeed raise concerns about the risks to spectators. However in 1996 such policies were a rarity, despite being recommended in the third edition of the 'Green Guide'. It was not until the 1995 FLA publication on 'Guidance Notes on Drawing up a Statement of Safety Policy for Spectators at Football Grounds' that, encouraged by the FLA Inspectors, those policy statements came to be formulated. In 2005 all 92 Premier and Football League Clubs have such a policy but the position in other sports remains very random – despite the importance of safety policies in all sports, not just football. The policy should show that, whilst responsibility for safety can be delegated, what cannot be delegated is the ultimate accountability. When things go wrong, this will rest with the club chairman or the event organiser. I suspect that few such persons really understand the significance of this accountability.

Frosdick touches upon the need for competent personnel to manage safety and in 1996 the Football Safety Officers' Association (FSOA) was still in its early stages of development. Since then the organisation's influence on safety at football grounds has grown along with the professional competencies of its members. What Frosdick overlooks was a similar growth in the skills and competencies of the FLA since it was formed in 1991 with a stated objective of being the lead authority on safety at sports grounds. Many, including myself, would not disagree that this objective has been clearly achieved. The consultancy gap between safety theory and practice, which Frosdick saw as an opportunity for his programme, has largely been filled by the FSOA and FLA.

Frosdick's suggestion of a consultancy to provide an independent audit and inspection service is therefore unnecessary, providing senior ground management, safety officers, local authority enforcing bodies and (in football) the FLA are properly discharging their spectator safety roles and responsibilities. The evidence from the FLA annual reports would indicate this is the case but of course this independent element of monitoring and inspection does not apply to sports other than football. That is why I would argue for the FLA's remit to be extended to all sports. The government have for several years signalled their intention to introduce primary legislation to bring this about, but the competition for parliamentary time has so far prevented progress. From my knowledge of other sports this extended FLA role is long overdue. I hope that we do not have to wait for a disaster in another sport before government recognise the need to expedite this change. For too long in the UK spectator safety has been governed by 'tombstone legislation' – we have waited for people to die before we legislate to prevent this.

I do not disagree that professional health and safety at work advice will generally be necessary, unless somebody in the organisation has completed health and safety training to an appropriate standard. However developments in spectator safety management training, such as the FSOA 'Event and Matchday Safety Management Course', and developments in football steward training (referred to in Parts II and V) would indicate that the risks related to safety at football grounds are being professionally addressed. I would suggest that in 1996 event hazard identification and risk assessment documentation were a rarity in football stadia, but in 2005 they are the norm. Monitoring of such documentation, coupled with the external audit of football grounds by local authorities and independent monitoring by the FLA, means that the independent element called for by Frosdick is very much in place in football.

Frosdick's article was however very much the forerunner to his book with Lynne Walley in 1997 'Sport and Safety Management' published by Butterworth-Heinemann. The book looks at safety at sports grounds in the round and provides excellent reference material on issues relating to risk, responsibilities and accountability. Despite being eight years old this book is of equal value in 2005 and should be essential reading for anyone interested in safety and security at sports grounds.

Frosdick's vision of Staffordshire University helping to provide solutions to safety management problems had little impact on the world of football. However Frosdick's personal contribution to the development of good safety management practices will be evident from the sum total of the articles in this book. In particular, his involvement between 1996 and 2005 in the production of the multi-media 'Training Package for Stewarding at Football Grounds' is testimony to how consultancy can bridge the gap between theory and practice. The training package has earned him wide recognition in all sports and he remains today a leading authority on risk management, safety management and stewards training.

In 2005, whilst the issues in the article remain the same, the way those issues have been tackled – particularly in football – shows considerable progress. Senior football ground management can demonstrate that they accept their safety responsibilities and no longer rely on the police to fill the void, which the police of course no longer want to fill. Competent and professional safety management and stewarding is now the norm and not the exception. Hazard identification, risk assessment and the management of risk are now everyday activities for football safety officers, but I cannot say this about other sports with the same confidence.

1.2 – VENUES OF EXTREMES: THE PAF APPROACH

The original citation for this article is: Frosdick, S. (1996) 'Venues of Extremes: The PAF Approach', *Stadium and Arena Management*, Volume 1 Issue 1, November 1996, pp. 26-30.

Steve Frosdick highlights the complexity and diversity of hazard perception in public assembly facilities management.

Stadia and arenas, increasing referred to as 'Public Assembly Facilities' (PAFs), are venues of extremes. On the one hand, they are the setting for some of the most exciting, enjoyable and often profitable events in the world. On the other, they have been the scene of some terrible disasters. In 1966, 300 people died in a stadium riot in Lima in Peru. In 1992, one person was crushed to death at a concert in South Korea. The same year, 13 people were killed when a temporary stand collapsed early in a football cup semi-final in Corsica. Most recently, in June 1996, fifteen football fans died when a wall collapsed in a crowd stampede during a World Cup match in Zambia. And I have not mentioned either Heysel in Belgium or the catalogue of disasters involving the deaths of hundreds of British football supporters. The many well-documented near misses, for example the collapse of a stand during a Pink Floyd concert at London's Earl's Court arena, are further evidence of the potential for PAFs disasters.

Dealing with safety hazards and the risks to which they give rise must therefore be seen as an important part of PAFs management. Operators have a moral and legal duty of care towards participants, performers, spectators and staff alike. In some countries, they have a statutory duty to deal with risks to the health and safety of people who might be affected by the operation of their facility. And following a number of high profile legal cases, there is growing awareness of the civil and even criminal liability of PAFs management in the event of a disaster precipitated by negligent preparations.

Thus the importance of risk assessment has been brought into ever sharper focus. Understandably, the emphasis has been on public safety hazards, and whilst these must be paramount, it can be shown that looking at them in isolation can create operational difficulties for PAFs management.

Different Perceptions

The problem is that managers are faced with four competing demands. Commercial pressures require them to optimise the commercial viability of the venue and its events. Spectator demands for excitement and enjoyment require credible events staged in comfortable surroundings. Regulatory and other requirements for safety and security must also be met, whilst any negative effects which the venue and event may have on the outside world must be kept to a minimum.

Each of these areas contains sources of hazards and risks. My own research into strategic risk management in PAFs in Britain has shown how risk means different things to different people in different contexts. Regulatory perceptions of risk as breaking safety rules are predominant, and a multitude of agencies - emergency services, local authorities, governing and enforcement bodies - are involved in safety and security management. PAFs owners and operators are more entrepreneurial and give priority to commercial risks such as access control, pirate merchandise, ticket touting, cash handling and ambush marketing. Spectator and local residents pressure groups are more concerned with quality and environmental risks. The vociferous minority of spectators demand the right to sit (or stand) where they choose to watch the event as they please without being commercially exploited or having their enjoyment intruded upon by petty-minded officialdom. Local residents voice concerns about the impact of noise, litter, traffic, vandalism and parking. The majority of spectators tend to shrug their shoulders with a fatalist acceptance of the various hazards they endure as a result of the commercial, regulatory, and behavioural excesses of all the others.

Operational Conflicts

Successful PAFs management means striking an appropriate balance between these demands. Yet there are many examples of the operational problems which have arisen as a result of management failure to achieve this balance.

At a Newcastle United versus Sunderland football match in 1993, disorder broke out as a result of stewards and the police removing a Sunderland banner draped over a pitch-side advertising board. The banner was preventing the sponsor's name from getting television exposure. Perceiving the hazard, the commercial manager had deployed the stewards without consulting anybody. The police had pitched in to help the stewards when they perceived a public order hazard. Two officers snatched the banner and a fight broke out. I was watching from the control room with the stadium safety officer: he was furious at the safety hazard created as members of the crowd suffered the results of the fight.

A second example comes from 1995 from a football ground in the north-west of England. Part way through briefing the senior stewards, the safety officer was called away to speak urgently to the commercial manager. The latter told him there was a fire in a hospitality suite and requested his immediate attendance. The commercial manager was teasing but 'I thought that would get you here quick'. In fact he wanted the pitch covers to be moved from where they had been folded up because they were preventing the advertising hoardings being seen. In dealing with the commercial risk, the manager thought nothing of disrupting the essential briefing for the senior stewards, which had to be curtailed. What would any subsequent inquiry into a real fire have made of the disruption and the irresponsible lie? And the pitch covers never did get moved!

These examples illustrate operational conflicts between safety and commercialism. Of course there are other conflicts, for example between safety and enjoyment, such as the indiscriminate banning at some venues of the flags and drums which add so much to the carnival atmosphere at sporting events. But the two examples from my own experience must suffice to make the point.

A Strategic and Systems Approach

My argument, then, is that to balance these differing demands and perceptions, the PAFs manager needs to adopt a more strategic and holistic approach to hazard and risk assessment. What this means in practice is an acceptance that nobody is wrong either to perceive a particular issue as a hazard, or to evaluate a risk in a particular way. It is therefore important to ensure that a broad range of perspectives are adequately represented in any risk assessment exercise. This is best accomplished in two ways. Firstly, any exercise should be undertaken by a group of people, rather than just one or two. Secondly, representatives of each of the four groups: commercial, regulatory, spectator and local resident, should be identified and invited to participate.

To illustrate the potential richness and diversity of this approach, I want to look at PAFs as systems, broken down into zones, such as is shown in Figure 1.2.1.

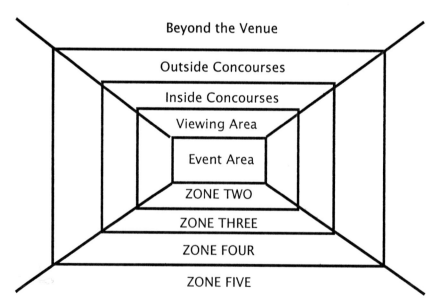

Figure 1.2.1 The stadium as a system of zones.

My site visits and analysis of briefing documents, contingency plans and training material has enabled me to catalogue, probably not exhaustively, the variety of hazards which have been perceived in each part of the system. The constraints of space mean that I shall concentrate here on the most commonly perceived hazards in just one zone, namely the area where the event is held (the pitch, track, or court, etc.) and the perimeter between it and the viewing zone.

Event Area and Perimeter Hazards

The Commercial Perspective

Threats to the interests of advertisers and sponsors form the principal sources of event area hazards perceived from a commercial perspective. The two examples cited above show the commercial importance of ensuring that perimeter advertising is clearly visible to the television camera.

Since accredited sponsors will have paid substantial fees to be associated with the event, there is also a perceived need to prevent 'ambush marketing' by other brands. At the Portugal versus Turkey match in Nottingham during the recent European Football Championships, several banners advertising Portuguese products were brought into the stadium and displayed whenever play and thus the cameras went in their direction. Stewards had to be more active dealing with these banners than they did with the well behaved crowd!

Television companies also pay handsomely for their access and are inevitably anxious to minimise the risks of high installation costs and poor broadcast quality in their choice of camera positions and cable runs around the event area perimeter. Conflicts arise when these choices create trip hazards or obstruct spectator sightlines.

The Spectator Perspective

Since their main purpose is to watch the event, any deficiencies in sightlines, in the physical event area and in the event itself provide sources of hazard to the enjoyment of the spectators. Restricted views arise from old PAFs designs, with roof props and even floodlight pylons around the perimeter of the event area. Unusually high perimeter hoardings, cage-type fences, inappropriately sited television cameras or excessive deployments of police/stewards around the perimeter represent further sources of hazards to sightlines.

Sports events may either become a farce or else be unplayable if surfaces, particularly grass, become too wet. The high jump section of the women's pentathlon competition at the Atlanta Olympics, where standing water was not properly cleared from the runway, adversely affecting the athletes' performances, provides an example. Enjoyment may also be threatened by a lack of credibility in the event itself. In boxing, a number of 'big fights' have ended in the first round because of the mis-matching of opponents. The early dismissal of a star player, even if justified, denies spectators the chance of enjoying that player's skills and may lead to their team adopting boring defensive tactics for the remainder of the match.

The External Disruption Perspective

Whilst it is clearly the zone beyond the venue which provides most hazards perceived by the outside world, nevertheless the event area itself provides two main sources. First is the noise created by the participants or performers. This is a particular issue with music events staged in stadia, where the sound travels beyond the stadium through the open air.

Second is the threat of articles from the event area being projected beyond the facility. Cricket balls 'hit for six' out of the ground can cause damage to property or injury to passers-by. Pyrotechnics set off on the event area provide a further source of hazard. During an early satellite television broadcast from a football ground in South London, some of the pre-match fireworks landed, still burning, on the forecourt of a petrol station down the street!

The Safety/Security Perspective

The principal sources of hazards from this perspective involve perimeter obstructions and the potential for adverse interaction between spectators and participants in the event.

Television cables, perimeter hoardings, fences and gates all provide tripping or obstruction hazards which may delay spectator egress onto the event area in emergency evacuations. The 1989 Hillsborough stadium disaster in which 95 football fans were crushed to death against a perimeter fence provides an extreme example.

Incursions onto the event area are the other principal concern. The perceived hazards range from attacks on officials or players, for example the on-court stabbing of tennis star Monica Seles in Germany, to damage to the event area, such as at Wembley Stadium after a notorious England versus Scotland football match in the 1980s. Conversely, participants leaving the event area cause similar concerns. Players who run into the crowd to celebrate goals or points scored frequently cause the crowd to surge towards them. And who could forget the pictures of Manchester United's Eric Cantona leaping into a stand to karate kick a spectator? Team benches provide a source of similar hazards, either because spectators misbehave towards them or vice versa.

Finally, we have the health and safety hazards which the event, the event area or the perimeter pose to the participants themselves. For safety reasons, the English football team nearly refused to play on a poor surface in China earlier this summer [1996]. And several boxers have died in the ring.

Looking at the other four zones would illustrate the different perspectives in further detail, and I hope to cover these in future articles. But from the event area alone, it can be seen that strategic risk assessment requires a broader focus than safety and security alone.

1.3 – VENUES OF EXTREMES: THE INNER ZONES

The original citation for this article is: Frosdick, S. (1997) 'Venues of Extremes: The Inner Zones', *Stadium and Arena Management*, Volume 1 Issue 2, February 1997, pp. 34-36.

Managing risk is a complex business. Steve Frosdick continues his analysis of the competing demands faced by public assembly facilities management.

In the launch issue of *Stadium and Arena Management*, I argued that managing hazards and risks was an integral part of facilities management. Public Assembly Facilities (PAFs) managers are faced with four different groups, each giving priority to different types of hazards:

- owners and operators, who are concerned with threats to their revenue streams;
- spectators, who wish to view and enjoy events staged in comfortable surroundings;
- regulatory agencies, who seek to enforce safety and security rules; and
- the community, who want their environment disrupted as little as possible.

Operational conflicts arise when managers fail to recognise these competing demands and are unable to achieve the right balance between them.

I suggested that PAFs managers needed to take a more strategic and systematic approach to risk assessment, and that one way to do this was to look at the venue as a zonal system. I looked at the event area and its perimeter with the viewing accommodation. I drew out the hazards perceived from each of the four perspectives and showed how these were indeed very different from each other.

In this second part, I want to continue by looking at the next two zones, taken together. These are the viewing accommodation and inside concourse zones, including the various technological systems used to support their management.

The Commercial Perspective

Design and fitting out are very much shaped by commercial risk concerns. Developers will want to recover as high a percentage of their capital costs as possible through advance sales of executive boxes and term tickets for premium seats. Operators will want to maximise the revenue streams from ordinary ticket sales by fitting as many seats into the facility as the various constraints will allow. And funding for many redevelopments has been underpinned by the idea of diversifying the uses made of the facility, through conferences and banqueting, on days when there are no spectators in to view an event.

Factors which interfere with the opportunity for spectators to gain access to the event represent a further source of risk. Venues want to sell as many tickets as they can, yet police insistence on 'all-ticket' matches or their refusal to allow sales on the day have adversely affected attendances at some British football grounds.

Access to the event – yes – but free admission – no. Revenue protection means that entry to the inside concourses needs to be strictly controlled. This ensures that only those who have either paid or else been properly accredited are permitted to enter the facility.

Having got the audience in, merchandising seeks to address the risk that ancillary spend per head will not be optimised. More and better retail outlets, together with branded confectionery and catering items, increasingly seem to provide the answer here.

Risk may also arise from anything which increases costs or which prevents the maximisation of promotional opportunities. There are frequent tensions here between commerce and safety/security. Commercial managers will want the level of security personnel employed to be no more than is necessary to deal with the numbers and type of crowd expected, whilst regulatory agencies will be tempted to up the staffing levels 'just in case'. Commercial managers will want to earn revenue by allowing access to promotions, yet the promotional activity may itself compromise safety. At an old London football ground, a local publisher was allowed to place a free copy of his newspaper on every seat in the main stand, which happened to be made of wood.

Finally, there are concerns not to offend the occupants of executive boxes and damage repeat business by over-controlling their behaviour. Thus normal security personnel may be replaced by 'lounge stewards' who are encouraged to show more tolerance and tact than would be the case with the ordinary public.

The Spectator Perspective

The main areas of spectator perceived risk concern the ease with which they can purchase the right to a seat/space and the quality of their enjoyment of the event.

Ticketing systems have become ever more sophisticated – indeed the launch issue of *Stadium and Arena Management* included a feature on Internet ticketing. Theatre box-offices are used to allowing the customer to choose exactly which seat they wish to purchase, but this is still a rarity in British sports. And there are still venues where even credit card sales over the telephone are not provided for and the prospective customer has to attend the venue in person to purchase the ticket.

Once they have gained access, spectators worry about whether they will be able to see the event. In addition to the viewing obstructions around the perimeter (mentioned in the launch issue), the view quality is affected by three factors: preferred viewing location, viewing distance and sightline. The preferred viewing location for athletics is the side where the finishing line is. Rugby fans prefer the sides whereas younger football fans prefer the ends. For most team events, optimum viewing distance is a radius of 90 metres whilst the accepted maximum is 150 metres. Yet in several famous stadia, most spectators are beyond the optimum and far too many are beyond the maximum. Sightlines are assessed using riser heights, tread depths and angles of rake. Ideally, the spectator should be able to see over the head of the person in front, but this has often not been achieved.

Given a decent view, the spectator is then concerned with enjoyment. Risks here arise from failures in maintenance – such as dirty or broken seats, from poor amenities and, above all, from being prevented from having a good time.

Spectator amenities will mainly be sited in the inside concourse areas. Here the fans are looking for both ready access and a choice of quality in catering and souvenirs, as well as for sufficient clean and decent toilet provision. All too often, they face the risk of their loyalty to the sport or team being unscrupulously exploited. Long queues, foul latrines and over-priced insipid fare are still the norm in too many venues.

Enjoying the event is clearly key, and it is here that spectators may unwittingly come into conflict with the regulators. For some sports fans, being forced to sit in a designated seat rather than to choose where or even whether to sit, is an infringement of the right to enjoy themselves. The same is true of regulatory restrictions on the banners, flags, air-horns, drums and instruments which go towards creating the carnivalesque atmosphere which so contributes to the enjoyment of the live event. Risk for spectators also arises when security personnel respond to their passionate partisan support and letting off steam as though it was hooligan behaviour.

The Safety/Security Perspective

A considerable proportion of the perceived risks arise and are addressed in the preparations for the event. Periodic inspections will be carried out on the structural integrity of the viewing accommodation, for example to check loadings, and extensive pre-event checks will take place to ensure that risk is reduced. There is a whole range of technological life safety systems – turnstile counting, crowd pressure sensors, lighting, closed circuit television, fire safety, communications and public address – which need to be working correctly to fully support the operational management of the event.

Furthermore, managers will want to be satisfied that the venue is clear of hazards and that all personnel are on post before they open the venue to the public. A recent under 21 international football match at Wolverhampton was delayed for nearly three hours after a suspect object was found during a pre-match search of the viewing zone. And several stadium events have been called off due to toilets of fire equipment having frozen.

In addition to its importance for revenue protection, access control is both a security and safety issue. Many venues are designed so that inside concourses and viewing accommodation together form self-contained areas, perhaps one for each side of the venue. Allowing too many people into an area creates a serious risk of overcrowding and possible disaster. Allowing unauthorised persons in compromises safety – if lots of people are involved – and security – if the person's intentions are sinister. At one London ground, a person walked through the players entrance dressed in a tracksuit and 'warmed up' with the teams until somebody realised he was an intruder.

Security risks arise from members of the crowd arguing over seat occupancy, committing criminal offences such as abusive chanting, throwing missiles, being drunk, fighting or reacting to the event with language or behaviour which is regarded as unacceptable by the authorities. Major crimes such as rioting, wounding or even unlawful killing are well known to have occurred in various sports. PAFs handle considerable sums of cash and several have been the victims of robberies.

Safety risks arise from areas of the viewing accommodation approaching capacity and from any factor which necessitates either a partial or total evacuation of the venue. These range from equipment failures – for example floodlighting – to fires, floods, gas leaks, explosions, bomb scares, structural collapses or serious public disorder.

Within the inside concourses, locked exit gates represent a particular safety hazard, since they prevent crowd egress in an emergency. Over fifty people burned to death in Bradford in 1985 and British football has learned this lesson. However the same cannot be said of all other sports or countries. In 1993, I went to a cricket test match where I found the exit gates to a wooden stand locked. I was horrified to be told that the steward who held the keys was taking tea in the pavilion several hundred yards away. In October 1996, I visited a French football ground where all the exit gates were kept padlocked yet unmanned throughout the match.

The External Disruption Perspective

Since it is the viewing accommodation and inside concourses which form the bulk of the PAFs structure, thus it is the impact of the built form itself which provides the main source of risk to the world beyond the facility. This is less of an issue where PAFs are constructed on greenfield or redundant industrial sites. However many PAFs are sited in cramped inner city locations and any redevelopment has to take account of the environmental impact on local residents. For example, the huge new stand at Dublin's Croke Park (the home of Gaelic football) was designed so that houses in the area would not lose sunlight either in the morning or the evening. And research carried out by Helen Rahilly at Liverpool University has shown how, to meet the considerable objections of local residents, the final design of Arsenal Football Club's North Stand was 'lower, lighter and far less bulky than the original plans had suggested'.

Noise and light pollution are further sources of risk. Light pollution occurs when the glare from floodlighting spills over onto surrounding properties, whilst noise pollution refers both to the noise of the crowd, which is perhaps unavoidable, and to the transmission of music and messages over public address systems which carry beyond the venue.

The constraints of space mean that I have highlighted only some of the principal areas of risk arising within each of the four perspectives. There are still considerable gaps in the analysis, which I acknowledge is skewed towards my own research in British football grounds. Many readers will undoubtedly be able to add considerably to what I have outlined.

But this is exactly my point. Having now looked in overview at three of the zones in the PAFs system, the full complexity of the balancing act required from facilities managers is becoming ever clearer. In a future issue of *Stadium and Arena Management*, I shall look at the two remaining zones and then seek to draw out the implications of the analysis, not only for management, but also for the PAFs design and construction industry. It may be that early recognition of the complexity and diversity of hazard perception could assist in designing out operational problems at the outset and thus save the considerable costs of retro-fit and over-staffing.

1.4 – DANGER, DISRUPTION, FINANCE AND FUN

The original citation for this article is: Frosdick, S. (1997) 'Danger, Disruption, Finance and Fun', *Stadium and Arena Management*, Volume 1 Issue 3, April 1997, pp. 34-36.

Steve Frosdick concludes his analysis of the complexity and diversity of hazard perception in public assembly facilities management.

In the first two issues of *Stadium and Arena Management*, I have been looking at Public Assembly Facilities (PAFs) as systems broken down into five zones: the event area, viewing accommodation, inside concourses, outside concourses and the neighbourhood beyond the venue. I have been drawing out the different types of hazards perceived in each zone by each of four different groups: commercial owners, spectators, regulatory agencies and the local community. I have suggested that PAFs managers need to recognise these different points of view and seek to strike an appropriate balance between them.

In this third and final part, I want to give a brief overview of the principal hazards perceived in the final two zones – the outside concourses and the neighbourhood – and then conclude by discussing some implications of the analysis for PAFs management.

Hazard Perspectives

Since merchandising makes a valuable contribution to revenue, thus pirate merchandise represents a serious commercial threat for PAFs managers. Not only will no license fee have been paid but the merchandise itself may be very poor, damaging customer perceptions of quality and thus reducing official sales. Within the UK, brands are protected by copyright legislation and a number of venues have employed security personnel to patrol the environs to seize any pirate merchandise. This is sometimes undertaken in conjunction with Trading Standards officers who have the power to prosecute offenders.

Similarly, as PAFs look to earn more revenue from sponsorship and licensing arrangements, so ambush marketing becomes an important commercial hazard. Writing in *Panstadia International* in Summer 1993, James M Curl described how ambushers can be 'locked-out' through the complete control of images in and around the venue. The environs can be controlled by local government permit whilst the outside concourses can be strictly patrolled to enforce restrictions on banners, signs and even clothing.

Spectators are rather less likely to perceive the wearing of a branded sweatshirt as a hazard. Outside the venue, their concerns centre around the ease with which they can gain access to the event and then get away afterwards. Perceived hazards arise from inadequate public transport, poor road capacities and difficulties in parking near the venue. Quite perversely, the same people who see their convenience threatened by parking restrictions before the event, will bemoan the absence of action against the parked cars which hinder the progress of traffic leaving after the event. Within the outside concourses, spectators wish to be guided effortlessly towards the right entrances and then gain admission without having to queue for more than a couple of minutes. Clear information on tickets, the best of signage and sufficient turnstiles or access points are essential if spectator frustrations are to be avoided. The needs of disabled patrons must also be met.

From a safety/security perspective, the principal hazards are perceived as disorder and overcrowding arising during the periods before and after the event. Particularly within football, there are public order concerns around spectators travelling to the match or gathering in public houses near the ground. Police have sophisticated intelligence systems which may cause them to believe there will be disorder and thus wish to make careful arrangements to provided supervised or even segregated routes to and from the stadium for the supporters of different teams.

The cramped inner city locations of many facilities create real risks of overcrowding in the surrounding streets, particularly as the start of the event approaches and many people are still outside waiting to gain admission. It is for this reason that the kick-off times of football matches are sometimes put back by 15 minutes and public address announcements made outside the ground to reassure fans they will not miss the match. Similar crowding risks arise where the venue is sold out and there are substantial numbers of people locked out of the event. Equally, at the end of the event, the exits onto the outside concourses will be opened in plenty of time and traffic stopped to ensure the crowds can disperse as freely as possible.

Within the outside concourses, supervised access control is perceived as essential to screen out drunken persons and minimise the risks of people taking dangerous articles such as flares, missiles or weapons inside the venue. For all-ticket capacity events, cordons may be placed at the boundary to the outside concourses, or even in the surrounding streets, to reduce crowding by restricting access to ticket holders only.

As far as local residents are concerned, the principal hazards arise from the nuisances created on the days when the venue is open for mass spectator events. These include the noise of the approaching crowd, the litter which the crowd leaves in its wake, minor disorder or vandalism, people urinating in the street or in front gardens, traffic congestion and the impossibility of finding a parking place near one's own house. Whilst these quality of life issues may be less keenly felt where venues are sited out of town or on derelict industrial land away from the main conurbation, nevertheless studies by geographers have shown how negative effects are experienced over quite some distance around a facility. Such widespread effects go some way to explaining the 'not in my back yard' campaigns so often mounted against PAFs developers seeking a site for a new facility.

Implications for PAFs Management

Over the course of the three articles in this series, we have seen how managing hazards and risks is an integral part of PAFs management. Four different groups give priority to different types of hazard in the five zones in the PAFs system. Commercial viability, excitement and enjoyment, safety and security and environmental impact are all legitimate perspectives on risk. Striking an appropriate balance between them presents PAFs managers with a major challenge.

So what I want to do now is suggest a practical management approach which acknowledges the validity of different perceptions and thus allows for greater richness, diversity and consensus in the analysis. The approach involves the five stages of identifying, estimating, evaluating, managing and monitoring risks.

Hazard identification and risk estimation can both be carried out at a full-day risk assessment workshop at which representatives of each of the four groups – commercial, regulatory, spectator and neighbourhood – can be invited to participate. The representatives should have a 'hands-on' knowledge of the venue and may therefore be relatively junior. Three representatives from each group would give a manageable workshop of twelve participants. Using the idea of the PAF as a system, with each zone broken down into a number of smaller areas, the group should be facilitated through the identification of any hazards which they perceive in each area. Participants should be assured that there are no right or wrong answers, that candour is welcomed and the validity of all views will be recognised. This facilitated process will result in a comprehensive list of perceived hazards. Each hazard should then be considered by the whole group and a collective judgement made about its probability of occurrence. A second collective judgement should then be made about the potential adverse consequences if the hazard did occur. I would suggest that there are four types of adverse consequences to consider: for the profitability of the business (including its exposure to liability); for the enjoyment of the spectators or participants; for public safety and order; and for the community and environment in the outside world.

Thus the outcome of the workshop is a comprehensive hazards register which should provide a substantial reference document to support the operational management of the venue. But PAFs managers cannot reasonably be expected to tackle all the hazards, nor will it be cost-effective to try to do so. So the risks will need to be evaluated by an appropriate forum, which may well be the board of directors of the company running the venue, or their public sector equivalents. The principle governing such evaluation is that risk should be reduced to a level which is 'as low as is reasonably practicable'. Hazards which have been judged to be of lower probability or consequences will be designated as 'low risk' and will be accepted as tolerable. At the other end of the scale, hazards which have been estimated as higher probability and consequences will be designated as 'high risk' and therefore subject to remedial action, in some cases irrespective of cost, through the preparation of appropriate risk management plans. Where the boundaries fall between these two categories will be a question of management judgement, and, once decided, will determine which hazards are designated 'medium risk'. These may require careful monitoring with action where something can be done at a cost less than the benefit of the risk reduction.

The identification, estimation and evaluation processes should then be reported as a formal risk assessment. This documents which hazards have been identified as priorities and how and why those decisions have been made. So the risk assessment process helps PAFs managers in discharging their accountability for profitable, safe, enjoyable and minimally disruptive facilities management.

For each high risk or relevant medium risk, a second workshop should meet to consider what action can be taken to control the risk and reduce the probability and/or consequences to a tolerable level. Again, the workshop should comprise representatives of each of the four different groups, although it would now be appropriate for more senior management to be involved in the exercise. The countermeasures defined, the resources assigned and the responsibilities allocated should be recorded. The outcome of the process will be the risk management plan. Having implemented the plan, regular monitoring is important to ensure that the implications of any changes are considered and appropriately acted upon. A formal review of the hazards, their estimated probabilities and consequences, tolerability and any risk management measures proposed, should therefore be carried out at appropriate intervals, for example after major building work. The hazards register, risk assessment and risk management plan should be amended as appropriate and reissued accordingly.

This approach can work not only for existing PAFs, but also for venues which are being renovated or even which have not yet been built. Engineering drawings can be used to determine the zones and areas in the proposed system and workshop representatives can be drawn from the groups who will eventually be involved in the facility. Carrying out a risk assessment should assist in early resolution of potential operating difficulties, thus allowing for the design to be changed to eliminate unnecessary operating costs, safety problems, external disruption and so on. For example, there may be a perceived hazard that personal radio communications will not work under a large stand because of the density of concrete. Identified at the design stage, this allows for leaky feeders to be built in during construction rather than as an expensive retro-fit. For the PAFs managers who will eventually be dealing with danger, disruption, finance and fun, early prevention is surely better than a cure.

1.2-1.4 – COMMENTARY BY JIM CHALMERS – A STRATEGIC APPROACH TO RISK ASSESSMENT

This series of articles written by Frosdick in 1996 and 1997 presents his strategic planning approach to the complexity and diversity of differing perceptions of risk in what he terms Public Assembly Facilities (PAFs). Of course he is advocating what he sees as best practice, but it needs to be acknowledged that his approach to risk assessment is only one of several that can be adopted by venue management. Similar advice can be found in Health and Safety Executive publications and in Section 2.3(e) of the 'Green Guide' to Safety at Sports Grounds. The articles therefore need to be read in conjunction with this other guidance.

The term 'Public Assembly Facility' was first used in 1995 by Wootton and Stevens from the Swansea Institute of Higher Education. Frosdick and Walley adopted it for their book 'Sport and Safety Management', but otherwise I have never heard it used in the industry. The term might therefore put practitioners off the articles. I would have preferred Frosdick to have stuck to what the articles are all about – namely sports stadia and arenas – without confusing readers with the term PAF.

Another problem with Frosdick's approach is that, whilst it is comprehensive and theoretically interesting for the student, it appears too bureaucratic and complicated to administer to appeal to the average stadium manager or safety officer. Frosdick is correct when he speaks of the various competing demands faced by venue management. He refers to commercial, spectator, regulatory and community perspectives, all of which have different perceptions of hazards and risks. But aggregating all the various interests up into four groups perhaps oversimplifies the story. For example within a football club virtually every employee of the 'commercial' concern has their own view on risk – and health and safety legislation encourages this. Club chairmen, directors, team managers, players, stadium managers, safety officers, crowd doctors, finance officers, ticket office staff, grounds maintenance staff and stewards are only examples of those whose diverse opinions have to be considered by the safety officer in the risk assessment process.

The 'regulatory' point of view includes the local authority, the emergency services, voluntary aid agencies and (in the case of football) the Football Licensing Authority. From a practical point of view I know from first hand experience how forceful all of these bodies can be in articulating their risk perspectives, not just concerning the venue but also their own employees. Frosdick also refers to 'spectator' and 'community' perceptions, but there is a real dilemma as to who actually represents the many thousands who make up the spectator audiences and local communities. Thus the reality of managing diverse risk perspectives is more difficult than the somewhat clinical image presented by Frosdick in his articles. Ongoing consultation by the safety officer rather than a one-off workshop is in my view the key to understanding everyone's point of view.

Frosdick explains how a venue can be broken down into five zones, although every sporting venue is different and in many small grounds zones three and four (the concourses) do not exist. The four competing demands are examined in respect of each of the five zones, providing a framework of 20 categories against which to identify hazards. This approach has much to commend it since it allows the safety officer to analyse risks in bite size chunks, rather than thinking of the venue as a whole.

Nevertheless, most safety officers have neither the time nor the resources to operate the types of processes suggested by Frosdick. These may be appropriate as a strategic one-off, perhaps associated with designing the operational procedures for a brand new stadium. However, my experience is that most safety officers try to keep the operational risk assessment process as simple as they can. Their job is to meet their duty of care by complying with regulatory requirements for a suitable and sufficient risk assessment. I would not want to see them paralysing their operation through excessive analysis and bureaucracy.

That said, hazard identification and safety risk assessment are dynamic processes. They have to be updated for every event, rather than treated as a generic one-off, the results of which will last all season. In the case of football matches these processes require information and intelligence on the numbers of home and visiting fans, the potential for disorder, methods of travel, other events in the area, the various stadium activities and the weather conditions. I am not clear how often Frosdick thinks his suggested risk assessment workshop sessions should be held, but in reality the outcome of what he proposes occurs week in week out at football clubs.

For example at Kidderminster Harriers FC, prior to each event, there is a meeting of all heads of departments, including the commercial side, to discuss the details of the event. From this the safety officer will start the hazard identification and risk assessment process. In a pre-event meeting with the local police both spectator and community issues are discussed and this detail will be used to update the risk assessment documentation. I would suggest that at most clubs the issues which concern the community are generally well known. In our case the main issue is one of spectator cars parking in residential streets and restricting access by residents. Complaints received from spectators at previous games are also part of the risk assessment process. Having taken all the information on board the risk assessment documentation is completed but is not finalised until the morning of the event. The risk assessment process does however continue throughout the event and if any incident occurs the safety officer will document and record this as part of a dynamic ongoing risk assessment process. All of this is done on a weekly basis without the need for full-day risk assessment workshops.

Frosdick does not mention the Football Safety Advisory Group (SAG) meetings chaired by the local authority. This group meets between two and four times a year and includes representatives from the club, emergency services and Football

Licensing Authority. Some groups also include supporters' representatives and locally elected councillors. Their role is to advise the local authority of stadium or spectator related issues which may affect the safe capacity of the stadium. In addition every Club in the Premier and Football Leagues must hold at least one exercise a year to test and validate the contingency plans for the venue. I consider that, taken together, the event planning meetings, dynamic risk assessment process, the SAG and the annual exercises deliver more value than the risk assessment workshop advocated by Frosdick.

I was interested to read Frosdick's views that the risk evaluation decision should be carried out by the board of directors running the venue or their public sector equivalents. In the world of football, chairmen and directors recognise that risk assessment can only be undertaken by trained and competent safety officers or stadium managers. The risks having been assessed, the question is then whether chairmen and directors are competent to evaluate whether the organisation should accept a particular risk or not. Given the complexity of safety management, I would question the competency of the average director to participate in this process without appropriate training.

Since the articles were written, knowledge, understanding and training in hazard identification and risk assessment in sports grounds have moved on. Much of what Frosdick wrote about in 1996 and 1997 is now standard practice. His suggestion of four categories of risk in five zones of the stadium provides a useful framework, but his four categories in my view oversimplify the complex range of perspectives which need to be captured through ongoing consultation. His suggestion of using workshop processes to assess risks is too time-consuming and bureaucratic to appeal to operational safety officers, who in any event achieve what Frosdick is advocating through their dynamic ongoing risk assessment processes.

1.5 – CONSTRUCTING IN TIGHT CORNERS

The original citation for this article is: Frosdick, S. (2001) 'Constructing in Tight Corners', *Stadium and Arena Management*, June 2001, pp. 18-20.

A multi-million pound redevelopment is well underway at West Ham United's Boleyn Ground in East London. Steve Frosdick describes the project.

A good number of UK stadia are in cramped inner city locations. Many of these venues have faced the problem of redevelopment within a very constrained site. The Millennium Stadium in Cardiff, butting onto the River Taff, was rebuilt with a tolerance of less than five centimetres around its perimeter. Liverpool's Anfield Road stand had only the footprint of the stand itself available as the building site. Clubs such as Manchester United and Aston Villa have built new stands on stilts over roadways, whilst Newcastle United and Liverpool have both built upper tiers as completely separate structures over and above existing stands. Blackburn Rovers were able to create space by purchasing and demolishing an adjacent factory and houses, whilst Arsenal are so landlocked that they can only add capacity by relocating altogether.

West Ham United have also had to address the question of adding capacity and other facilities within a tight site. The unique West Ham solution involves moving forwards into a diversified facility by moving one stand backwards and shifting the pitch and end stands sideways.

John Ball has been Stadium Manager and Safety Officer since shortly after the 1989 Hillsborough disaster – the catalyst for radical change in UK stadia. He inherited a ground with two large covered terraced ends, a main stand built in 1925 and a smaller side stand with a single tier of seating, below which was the famous 'chicken run' terrace. Together with his assistant, Ron Pearce, John has overseen the redevelopment over twelve years to date. The long term project has been driven by a range of risk management concerns affecting the commercial business, the spectator experience, safety/security and the local community.

The UK government safety requirements for the elimination of standing accommodation at many football grounds necessitated the demolition of the two end terraces. The north terrace – previously the home end – was a solid banked structure with only one route in and out. This made demolition difficult. The new north Centenary Stand is a two-tier structure, with the lower away zone topped by a second tier family area. This is good safety practice since at least on club has made the mistake of putting the boisterous sections of the home and away support in different tiers of the same end stand. The south terrace was concrete over a void and the club were able to buy some land to its rear and replace it with the new home end – the impressive Bobby Moore Stand. It was also farewell to the 'chicken run' with its terraced paddock replaced by seating.

Notwithstanding its splendid all-seated status, West Ham's return to the Premier League in the 1990s highlighted two substantial commercial pressures. First, the demand for tickets was such that the 27,000 odd seats were sold out almost every match. Adding capacity was essential if the club was to maximise ticketing revenues. More significantly, the club's catering and hospitality facilities were very limited. There were only a few executive boxes available holding 200 people – all in the Bobby Moore Stand – and the three small function rooms held a total of only 270 people. As John Ball explains, 'For commercial reasons, the club had to transform from a 20 or so days a year football operation into a seven days a week business'. So adding diversified facilities was just as important as adding capacity.

Restricted views for spectators were also a problem. The main west stand was 75 years old and most of the seats in the upper tier suffered some obstruction from the roof pillars. In the lower tier, views from seats towards the rear were quite severely restricted by the pillars supporting the upper tier. There was also some obstruction from roof supports in the east stand. Such restricted views are no longer acceptable, nor are they compatible with the ticket prices charged for Premier League football.

The lack of capacity, the absence of hospitality facilities and the restricted views all meant that the two side stands had to be replaced, but the tight site posed a problem. The east stand was sandwiched between the pitch and the street outside and its footprint was four metres too narrow for the two-tier stand required. So could the pitch be moved across to create the extra space?

In the area between the main road, the side of the north (Centenary) stand and the back of the west (main) stand were a secondary school, a car park and then a junior school. The secondary school had a falling number of pupils and, when it was earmarked for closure, the club were able to buy the site and demolish the school. In the space created, the club built a brand new £2.5 million junior school. A land swap then took place and the junior school relocated from behind the west stand to the side of the north stand. The old junior school was then demolished, leaving the club with a good-sized site behind the existing west stand.

Careful planning was now needed to ensure minimum loss of ticket revenue, minimal disruption to the community and the maintenance of all safety requirements during the next stage of the development. It was also vital to preserve the renowned intimate atmosphere of the Boleyn Ground, with the front rows of fans sitting just a few feet from the pitch. The ingenious and unique solution was to build the new west stand behind the existing one, even whilst the latter was still in use. A design brief was produced and an invitation to tender for a design build contract issued. Taylor Woodrow, who had previously built Derby County's excellent Pride Park, were the successful bidders and work began in June 2000.

Building has continued with no loss of existing capacity and the current west stand was demolished in May 2001, immediately following the final game of the season. The upper tier of the new Doctor Martens Stand – named after the footwear manufacturer and club main sponsor – will be open in time for the start of next season. Such is the increase in capacity that all the west stand season ticket holders will be able to be accommodated in the new upper tier alone, and planning is in progress to offer each person a seat in a geographically similar position to their current place. The lower tier and fitting out of the additional facilities will be completed during the course of subsequent weeks.

The new Doctor Martens Stand is a conventional cantilever construction and will wrap round the corners to provide side extensions to the existing end stands. It will have two tiers, separated by two rows of executive boxes with space for 760 guests. The boxes are fronted by viewing galleries and, unusually, will be reconfigured as quality hotel bedrooms on non-match days. As well as the hotel, there will be club seats, a 400 seater restaurant, other function rooms, two classrooms for school and community use and a new stadium control facility. This will all amount to a massive increase in the catering and hospitality operation and allow for the ground to become genuinely multi-purpose.

The main problem during the development has been the need to provide safe access through a changing building site for west stand patrons on match day. This has been achieved by clearing a safe passage of sufficient width through the site, removing all materials and plant from the passage and erecting protective hoardings. The position of the passage has varied for each match but has been subject to an inspection and safety certification process by the local authority on a match by match basis. It has also been necessary to install cable television facilities for the few residents whose aerials were affected by the bulk of the new structure. However, according to John Ball, 'We had to do a lot of community liaison during the building of the Bobby Moore Stand but this is a bigger, more accessible site. There is less impact on residents with this stand and we have quickly sorted out any problems'.

The space created by the demolition of the old west stand will allow for the pitch to be moved 15 metres sideways, which in turn will create the space needed to redevelop the east stand. This will also wrap round into the sides of the existing end stands, the east corners of which will therefore need to be partially reconstructed. However the pitch will not be moved immediately so, as John Ball puts it, 'The ground may look a bit lopsided for the start of next season, although we are hoping to move the pitch over a little and centre it up if we can'. Whilst there might be some loss of proximity to the pitch during the works, the end result of a 40,000 seater diversified facility should prove to be really worthwhile.

1.5 – COMMENTARY BY JIM CHALMERS – STADIUM REDEVELOPMENT AT WEST HAM UNITED

The article written by Frosdick in 2001 provides a useful case study of the issues faced in the strategic management of change created by a major stadium redevelopment. The case study examines the redevelopment of the Boleyn Ground, home of West Ham United FC. The article is all the more interesting since the developments are viewed through the eyes of the club safety officer and the problems he had to face in ensuring the stadium was open for business as normal on a match day – despite being a building site at the same time. The article has particular significance for those safety officers and stadium managers who are undergoing a stadium redevelopment programme. It certainly reminds me of my involvement in stadium developments as an Inspector with the Football Licensing Authority from 1991 to 2003.

A brand new stadium development involving a new site has its own particular planning and construction issues, and I experienced these with new stadia built for Northampton Town FC and Oxford United FC. But one problem a new site does not have is the playing of fixtures at the same time as major reconstruction works are taking place. This is the more common experience and in my case involved redevelopments at Aston Villa FC (three stands), Birmingham City FC (three stands), West Bromwich Albion FC (three stands), Coventry City FC (three stands), Wolverhampton Wanderers FC (four stands), Walsall FC (one stand), Cheltenham Town FC (one stand), Kidderminster Harriers FC (two stands) and Wycombe Wanderers FC (two stands). These are only a snapshot of over 200 new stands built in England and Wales in the post Hillsborough era.

In all these developments the commercial enterprise issues raised by Frosdick were very much part of the planning and design process. Whilst replacing old standing terraces with seats was the catalyst for change, it was also an opportunity for clubs to re-examine the commercial operation of their stadia. Most Clubs, like West Ham United FC, recognise that they cannot exist on football-related revenue alone. Stadium usage has had to be diversified to maximise income. That is why even small Clubs such as ours at Kidderminster Harriers FC spend so much time and effort in promoting the use of our executive, conference and catering facilities which are available seven days a week. Weddings are now commonplace at suitably licensed stadia with the bride and groom walking down the centre tunnel as distinct from down the aisle.

This wind of change in realising the commercial value of stadium can also be judged from the number of chairmen and directors who have been appointed, not because of their knowledge of football, but for their business and entrepreneurial skills. As commercial activities have grown, so has the number of chief executives and commercial managers needed to promote and manage the business opportunities. Thus in line with the physical changes to the stadium structures, there have been significant changes to the commercial operation of the game which some would argue have not always been in the best interest of the sport or the fan.

However as Frosdick's article clearly explains the real pressures comes in converting a fully operational stadium into a building site and then back again to an operational venue, all in the space of a few days and throughout an entire football season or seasons. The case study examines how this was done at West Ham United but it really does not do justice to the degree of planning needed for the venue to be able to open. I know from personal experience that on the day before and on the day of the event there would have been detailed meetings between the safety officer, other club officials, the local authority, the police, the fire service and the construction company, to finalise when work on site would be completed, the security of the site and the state of readiness for all other parts of the stadium likely to be affected by the development. A risk assessment process supported all of this and unless the local authority was satisfied that the venue was reasonably safe for the admission of spectators then the stadium would not have been allowed to open. I recall on one occasion this approval was not given until 1.25 pm – and the turnstiles opened for business five minutes later. In this regard the article fails to emphasise the safety officer's time and effort to produce the necessary hazard and risk assessment documentation.

Despite all the precautions taken – and some are identified in the article – accidents can still happen. In one case, half way through a game I attended in the Premier League, a steel girder fell from the new roof and crashed onto the empty seating deck below. The noise was horrendous and, for one horrible moment, the safety officer, thousands of fans and I thought the whole steelwork was about to collapse. In another case at another ground part of an external scaffolding collapsed onto an access route. Thankfully this was empty of spectators at the time. Just one hour later this would have been the route used by fans to access the turnstiles.

The provision of site security was a major undertaking, particularly on high profile or all-ticket games, when security staff with dogs were needed to deter fans from trying to access the ground via the building site. There is however no limit to the ingenuity of football fans and at one Premier League game I noticed what appeared to be twelve workmen sitting in a partially built stand having a superb view of the game. Since it had been agreed that no workmen would be on the site, the safety officer sent stewards to investigate. It turned out that a dozen fans had found a hole in the outer security fence and had 'borrowed' a dozen protective hats from the empty site office. What if this had been on the day the steel girder fell? Just imagine the implications for the club and construction company had anyone been killed or seriously injured.

Frosdick also refers to the important issue of community involvement in major stadium redevelopments. With so many stadia located in urban and inner city areas, they are surrounded by residential properties which can be seriously affected by any new structure. The article provides a practical demonstration of how West

Ham United FC resolved the problems of local television interference. I cannot honestly say that all football clubs have been such good neighbours. Whilst stands have to meet planning and building regulations, they should also not interfere unduly with the quality of life of those who have to live in their shadow. In my experience, clubs have been able to show they appreciate the concerns of local residents by inviting them to planning and development meetings, providing a complaints hotline, regular cleaning of streets used by construction vehicles, limiting weekend and evening working on the site and by inviting residents to view the facility when completed. The effect on the community should never be underestimated and the practical help given by West Ham United FC in the case study was an excellent public relations exercise.

Presented as a practical case study, based on the vast experience and knowledge of the club safety officer, the article remains an excellent case study for any stadium manager or safety officer about to undergo their first major stadium development.

PART II – SAFETY MANAGEMENT

Introduction

Part II comprises four articles which deal with issues relating to the safe management of a sports ground. The articles look at how safety management has developed over the years, examining various risks and aids to the management of spectators such as the use of Closed Circuit Television (CCTV) and the police use of CS spray in a sports ground. Finally, Part II suggests how the engineering risk management technique of 'Hazard and Operability Study' (HAZOPS) could be used to integrate design and management in a sports ground.

2.1 – The Development of Football Safety Management

Writing in 1996, Steve Frosdick and John Sidney examined the development of safety management in football grounds, particularly the evolution of the Football Safety Officers' Association. This historical perspective has stood the test of time and is brought up to date in 2005 with comment on how safety management has been developed in other sports.

2.2 – CCTV Requirements Specification

In 1996 Steve Frosdick examined the growth in CCTV subsequent to the 1985 Bradford fire disaster in 1985. Many of the fears discussed could still be valid in 2005. However the more detailed and refined guidance now available, together with the development of modern CCTV technology, suggest that Frosdick's concerns have been overtaken by current practice.

2.3 – Police CS Spray and Crowd Safety

In 1997 Frosdick raised the possible threat to spectator safety should a police officer use CS spray in the confines of a sports ground. The article coincided with the introduction of CS spray as part of a police officers' personal protective equipment. Football Safety Officers and supporters alike shared his concern at the time. Over the years the concern about the police use of CS spray in sports grounds has not materialised. Comment is made about a disaster in Ghana in 2000 when the use of tear gas resulted in 123 deaths and about two incidents in the UK when police used CS spray in a sports grounds context.

2.4 – Case Studies in Stadium Design and Safety

These two case studies from 1995 examined how a lack of consultation by senior management in the design and operation of a stadium can lead to faulty design and subsequent management problems. Frosdick advocated the use of HAZOPS as a means of overcoming these difficulties. This was perhaps ahead of its time. Whilst there is no evidence that HAZOPS itself has ever been adopted in this context, nevertheless the principles have been embraced in many stadium design projects.

2.1 – SAFE CRACKING

The original citation for this article is: Frosdick, S. and Sidney, J. (1996) 'Safe Cracking', *Football Decision*, Issue 2, July 1996, pp. 48-51.

John Sidney and Steve Frosdick take an upbeat look at the safety management advances made in recent years.

British football grounds have been prominently represented in the tragic history of crowd- related disasters. The historical reasons for this are well documented. They include the former dilapidated facilities, general neglect of spectators' safety and comfort, hooliganism and poor leadership within the sport.

In terms of crowd management, the 1980s emphasis was firmly on public order. This was generally preserved through an increasingly repressive policing style, coupled with hard engineering solutions such as the high fences still seen in most continental grounds.

The 1989 Hillsborough disaster and the subsequent inquiry by Lord Justice Taylor have been widely recognised as the catalyst for major change in the British stadia industry. There was swift implementation of changes in planning, responsibilities, testing and improving the fabric of stadia, involving considerable energy and expense for clubs, local authorities, police and others.

Other key areas of change included the revision of the Guide to Safety at Sports Grounds (the 'Green Guide'), the scrapping of the proposed national membership scheme and the establishment of the Football Licensing Authority (FLA) and Football Stadia Advisory Design Council (now replaced by the Football Stadia Development Committee). New criminal offences of pitch invasion, racist chanting and missile throwing were also created.

Many of these changes involved changes in safety management. But the most notable and widely publicised change involved the elimination of standing accommodation at all FA Premier and Football League stadia, although the requirement was subsequently relaxed for the lower division clubs. There are only 115 designated sports grounds in England and Wales, yet the requirement for all seated accommodation gave rise to the overnight appearance of a thriving industry to service a massive multi-million pound programme of stadia redevelopment.

Much of the subsequent publicity concerning the standard of British stadia has continued to focus on the new and refurbished structures and the facilities provided within them. But the actual fabric of the stadium is only half the story. Equally important is the safety culture within the club and the quality of its safety management systems. Even the most modern ground could still be unsafe if the club cannot properly manage the safety of its paying customers.

But safety management has also been evolving. As Football Licensing Authority (FLA) Chief Executive John de Quidt recently pointed out, 'while the structure and fittings of the stadia have improved out of all recognition over the last few years, other equally important, though often less visible changes have also been taking place.' So what have the principal changes been and what benefits have they brought to spectator safety?

Football has undoubtedly become one of the most regulated activities there is. Local authorities are responsible for issuing the safety certificate specifying the stadium capacity and the conditions to be met before spectators are admitted, including the appointment of a safety officer and the provision of adequate stewarding. These certifying authorities, encouraged by the FLA, have become more rigorous in their safety certification and inspection procedures. Many have chosen to follow the modular certificate structure recommended by the FLA, although the contents have properly remained a matter for local determination. This has helped to ensure a balance between standardisation and taking account of differing local circumstances. The FLA have matured in their role through the issuing of licences to permit the club to admit spectators to the stadium, the monitoring of certifying authorities, the provision of advice and the carrying out of inspection visits. The Football Association (FA), FA Premier League and Football League have their own advisors who regularly visit grounds. And police, fire and ambulance services personnel are regularly in attendance at some matches.

Furthermore, a large number of different government agencies and other bodies have developed their own detailed guidance on managing aspects of public safety at sports grounds. The range of advice and recommendations is complex and clearly illustrates the growing importance of the subject.

Each Certifying Authority also chairs a multi-agency Advisory Group which meets periodically to discuss relevant safety issues. This allows for regular dialogue between the many agencies involved and facilitates improved operational co-operation. There has been a growing realisation that the different agencies need to work in partnership in preparing their respective systems and procedures. At many grounds, for example, it has been possible for the various agencies to agree joint contingency plans to deal with emergency situations.

Many of these groups also include a supporters' representative, ensuring that the fans' voice is also heard. This is important, because the supporters are often concerned to ensure that their enjoyment is not spoiled either by being commercially exploited by the club or by being subject to excessive restrictions by the police and stewards.

For reasons of both policy and costs, the police have sought to reduce their manpower commitments at public events including football matches. In the view of the Home Affairs Committee of the House of Commons and many others, 'higher profile stewarding supported by lower profile policing' represented the way forward. Following the Taylor recommendations, each club agreed a statement of intent with the local police, setting out their respective responsibilities for safety management. The general responsibility for safety began to be assumed by the club whilst the police role shifted to concentrate on crime, disorder and major emergencies.

This trend was reinforced by the authorities repeatedly pointing out that the responsibility for the safety of spectators lay with the Chairman and Directors of the club concerned. Football clubs began to recognise their legal and moral responsibilities for ensuring the safety of their customers and also to seek to reduce the escalating costs of special services of police.

These trends created the need for the appointment of club safety officers who would take operational responsibility for safety management. Initially, some of these appointments were on a part-time basis involving responsibility only for the match day operation and pre- and post-match inspections. But the majority of clubs began to realise that it was not possible to create an artificial distinction between crowd safety on match day and the health and safety of players, club staff and the public both on match day and throughout the week. Thus more and more safety officers became appointed as full-time safety professionals with a wide portfolio of responsibility.

As the police refocused on public order, so their intelligence systems for dealing with hooliganism became more sophisticated. The police appointed liaison officers for each club, creating a national network of local intelligence officers, co-ordinated by the Football Unit of the National Criminal Intelligence Service. Prior to any match, the liaison officer could call for an intelligence report, including full details and photographs of known trouble-makers, to assist in assessing the likely risk of disorder. This intelligence system was reinforced by the attendance of police 'spotters' representing the away club at appropriate matches.

These many changes in management practice were supported by widespread investment in modern safety facilities and equipment. Reflecting the growing trend towards multi-agency partnership, many of the newly built control rooms have included sufficient space and equipment for all the agencies involved in the operation. This has allowed for rapid consultation and co-ordination in emergency situations, allowing crises to be resolved at an early stage. The safety operation has been enhanced by the employment of technological life safety systems for access control, communications, emergency warnings, means of escape and surveillance monitoring. Under this latter heading, the installation of modern Closed Circuit Television (CCTV) systems is widely regarded as a key advance in allowing both for crowd safety monitoring and the identification of individuals engaged in criminal behaviour. And many grounds have properly equipped medical centres to provide for emergency aid.

One of the most significant developments since Hillsborough has been the formation and evolution of the Football Safety Officers' Association (FSOA). In the flurry of activity following the Taylor Reports, concerns were raised by both the football authorities and the practitioners themselves that standards of safety performance varied so greatly from ground to ground. Systems and practices had been evolving locally in a piecemeal and ad hoc fashion and there was very little uniformity. Spectators travelling away from home did not know whether they would be treated as an invading army, as valued customers or something in between the two. The FSOA was inaugurated on 29 October 1992 with the aims of improving safety at football grounds, promulgating best practice, enhancing the role of stewards and continually developing safety officers' expertise.

Full membership of the Association is open to safety officers and their deputies at grounds in the FA Premier and Football Leagues, whilst associate membership is open to any suitable person associated with the responsibility for safety at sports grounds. From an initial gathering of 28, the FSOA has grown, to date, into an organisation of 134 members. All professional football clubs except five are now represented, together with representatives of the football authorities, several local authorities, stewarding companies, major Scottish football clubs, rugby league clubs, consultants and academics.

And as its membership has grown, so the Association has become involved in a wider range of activities and become a more influential player in the stadia industry. It holds two national conferences each year at which members can debate safety issues and exchange good practice. These conferences are supplemented by regional meetings which allow for discussion of more local concerns. The FSOA is represented on the Advisory Group Against Racism and Intimidation (AGARI) and receives regular enquiries from companies wishing to promote their goods and services within the stadia industry. Most recently, Association members worked together with the football authorities to produce a sophisticated multi-media training package for stewarding at football grounds.

Whilst total safety can never be guaranteed, ensuring public safety means reducing the risks as far as reasonably is practicable. Taken together, these major management developments demonstrate the great strides made by football safety practitioners in seeking to deal effectively and professionally with the many complex safety hazards involved in stadium management.

Overall, British football grounds may properly now be regarded as enjoying the best standards of safety and comfort in Europe. These claims will no doubt be put under the microscope during the European Championships. The management and communications difficulties inherent in such a large one-off event should not be under-estimated and it is almost inevitable that operational problems will occur. Nevertheless, the experiences gained in recent years should ensure that the safety management systems are robust enough to cope.

However, as Lord Justice Taylor pointed out, safety is paramount and its enemy is complacency. And FSOA members are well aware that, notwithstanding the progress made since Hillsborough, there is much that remains to be done. Stewards' training is becoming more professional, but this is only the first rung of the training ladder. There is ongoing debate about training the trainers, assessment, the training of supervisors, and the professional qualifications of the safety officers themselves. The FSOA will continue to work with the football authorities in taking these and other safety matters forward.

2.1 – COMMENTARY BY JIM CHALMERS – THE DEVELOPMENT OF FOOTBALL SAFETY MANAGEMENT

This article was written in 1996, seven years after the Hillsborough stadium disaster. It provides a summary of the early development of football safety management – as distinct from stadium structural safety. The article focuses on the changing emphasis from police control of crowds to clubs managing crowds safely. That emphasis is still valid in 2005. However the practice of football safety management has advanced considerably in the intervening years.

The fourth edition of the 'Guide to Safety at Sports Grounds' (the 'Green Guide') was published in 1997 and for the first time had not been hurriedly revised as a result of a knee jerk reaction to a stadium disaster. Proper emphasis was now given to the importance of spectator safety management and it was by design, as distinct from accident, that chapter two of the 'Green Guide' is devoted entirely to 'Management Responsibility and Planning for Safety'. This early chapter clearly illustrates the importance of recognising that sports grounds safety extends far beyond the structures and installations themselves.

This principle provides the focus for the 2001 Football Licensing Authority guidance document on 'Safety Certification', aimed primarily at the local authorities which have to administer the safety at sports grounds legislation. Where there is doubt over the effectiveness of a sports ground's safety management, then the capacity should be reduced – and this has been done on occasion.

It could be argued that the most significant development has been the growth of the Football Safety Officers' Association (FSOA), both in membership and in influence in football. Other sports have followed this lead with similar associations in Scotland, in Northern Ireland and in rugby and cricket. The FSOA has a full-time administrator based in the East Midlands and the membership as at March 2005 was 252, with all Premier League and Football League clubs now represented. Associate membership is extended to football safety suppliers such as equipment installers and stewarding companies. For more information on the FSOA visit the website at http://www.fsoa.org.uk.

In 2002 the FSOA established a six day 'Event and Matchday Safety Management Course' for football safety officers, their deputies and assistants. By March 2005 over 100 safety practitioners had attended the course and its success can be judged by similar courses being adapted and run for safety officers in cricket, rugby and horseracing. There is even the prospect that the course may be extended into the training of safety personnel in European football grounds.

The original article referred to the importance of police intelligence in combating football hooliganism. In 2000 the FSOA developed its own website with part of the system devoted to a facility for uploading post-match reports. This allows safety officers to access and exchange information such as attendances, travelling support and any particular problems, including ejections and arrests. This information enables them to plan more accurately their events and use of resources, without relying solely on police intelligence. This capability can be further enhanced by clubs who participate in the 'SpecIntell' system of football intelligence gathering and the exchange of information on persons banned or ejected from grounds.

Despite the growth of the FSOA it is perhaps disappointing that only thirty safety officers or deputies have attained a Level 4 National Vocational Qualification (NVQ) in Spectator Control. Much emphasis has been placed on stewards attaining qualifications such as a Level 2 NVQ or the 'Football Stewarding Qualification' but there has been no emphasis on safety officers obtaining recognised qualifications. This is something the regulatory bodies must address and is perhaps the next major issue in the ongoing development of sports grounds safety management.

2.2 – WATCHING THE CROWD

The original citation for this article is: Frosdick, S. (1996) 'Watching the Crowd', *Football Management*, Volume 4 Issue 1, Winter 1996, pp. 28-30.

All Football Grounds are equipped with Closed Circuit Television Systems, funded by the Football Trust, to support the management of public safety. But how good are these CCTV systems in practice? Steve Frosdick investigates.

Following the disastrous fire at Bradford City football ground in 1985, the Football Trust made funds available to install Closed Circuit Television (CCTV) at the grounds of football clubs in the top two divisions of the then Football League. Some of the early systems were inadequate. Subsequently, some clubs installed improved equipment and most of the lower division clubs also applied to the Football Trust for funding for CCTV.

This was a new departure for both the football industry and the police. The CCTV systems then available were fairly basic, with poor quality monochrome images, and operators lacked training and expertise. Standards varied widely and, inevitably, some of the systems installed were deficient in both design and operation.

Nevertheless, as Lord Justice Taylor found in his reports into the 1989 Hillsborough stadium disaster, in which 95 Liverpool supporters were crushed to death, CCTV was perceived as having had 'a major impact on the hooligan problem inside football grounds as well as proving extremely useful to monitor safety measures'.

In his Interim Report into the Hillsborough disaster, Lord Justice Taylor made two recommendations concerned with CCTV:

Recommendation 21 – 'Closed circuit television should be so installed as to enable crowd densities outside the ground, within concourse areas and in pens and other standing areas, to be monitored before, throughout and at the end of a match.'

Recommendation 35 – 'There should be available in the police control room the results of all closed circuit television monitoring outside and inside the ground and the record of any electronic or mechanical counting of numbers at turnstiles or of numbers admitted to any area of the ground. Officers in the control room should be skilled in the interpretation and use of these data.'

In anticipation of the Taylor recommendations, the Association of Chief Police Officers (ACPO) Sub Committee on Hooliganism at Sporting Events had established a Working Party to examine the role of CCTV as an aid to public safety in football grounds. The Working Party decided on an appropriate image standard for CCTV being installed in accordance with recommendation 21. The Police Scientific Development Branch (PSDB) devised testing procedures to quantify this image and, together with the Working Party, produced a set of guidance notes for the procurement of CCTV installations for public safety at football grounds.

These guidance notes covered the assessing of the risk, establishing the operational requirement, specifying and testing the required performance level. They also dealt with tendering, selecting a contractor, commissioning, handover and maintenance. Regrettably, however, the extent to which the guidance notes were taken account of by football clubs and police forces was mixed.

CCTV systems have burgeoned in recent years in town centres, shops, banks and building societies as well as sports grounds. In response to concerns about value for money and the quality of image outputs, a Home Office Research and Development Project aimed at improving the effectiveness of CCTV safety and security systems was begun by PSDB in April 1992.

Acquiring a CCTV system has never been easier, However, as Jim Aldridge of PSDB has reported, 'acquiring one that works effectively and represents good value for money appears to be much harder'. The Football Trust have continued to fund new and upgraded CCTV systems for stadia. However, the PSDB research has highlighted concerns that customers, including police officers and football club personnel, may have been seduced by the technology and encouraged to make important technical decisions for which they are unqualified.

The 'CCTV Operational Requirements Manual' recently produced by PSDB outlines five different purposes for CCTV, and thus five different image qualities and content specifications for the systems. These are:

Monitoring – observers can determine the number, direction and speed of movement of people whose presence is known to them;

Detection – following an alert, an observer can ascertain whether or not a person is visible in the pictures displayed;

Recognition – viewers can say whether the person in the picture is someone they have seen before;

Identification – the identity of a previously unknown subject can be established beyond reasonable doubt; and

Other – for example, the reading of a car number plate.

Lord Justice Taylor's recommendations for CCTV were in respect of the *monitoring* purpose only. However, as Taylor indicated in his comments on football hooliganism, and as I know from my own research, CCTV systems in football grounds are variously used for all five purposes:

Monitoring – for signs of crowd distress, excessive crowd density, or outbreaks of disorder;

Detection – confirmation that stadium areas are clear of people;

Recognition – of known 'hard-core' hooligans or persons subject of exclusion orders;

Identification – of persons participating in criminal activity, for example serious disorder; and

Other – watching the match!

The identification purpose has assumed particular prominence as a result of two incidents in February 1995. These were the disturbances at the end of the Chelsea versus Millwall FA Cup Replay, which were investigated using CCTV footage; and the abandonment of the Ireland v England fixture in Dublin. The '10 point plan' subsequently announced by Glen Kirton, Director of the Euro 96 tournament, included the use of CCTV to allow the authorities to identify the occupant of each individual seat.

Critically, however, it is not clear to what extent the existing systems are capable of meeting these user requirements. One might reasonably suppose that some of these uses will not have been identified in the original system specifications.

A number of practical research questions therefore arise:

- For what purposes were the systems originally installed?
- What operational requirements were originally specified?
- What system of testing was originally specified?
- For what purposes have the installed systems been used in practice?
- Has the systems performance been adequate for these purposes?
- What standards are required?
- Are systems up to the standards required?
- What system of testing has been used to assess performance?
- Has any new or upgraded equipment been purchased to counter inadequate performance?
- For what purposes was this new/upgraded equipment installed?
- What operational requirements were specified?
- What system of testing was specified?
- For what purposes have the new/upgraded systems been used in practice?
- Has the systems performance been adequate for these purposes?

- What system of testing has been used to assess performance?
- Can the performance of the existing equipment be improved to meet the actual operational requirements?

These are important questions which, I believe, the football industry has not yet begun to sufficiently address. One would like to think that the stadia to be used for the forthcoming European Championships have no deficiencies in their CCTV systems, and this may well be the case.

But does anybody know for sure?

Figure 2.2.1 Control room at The Valley, home of Charlton Athletic.

2.2 – COMMENTARY BY JIM CHALMERS – CCTV REQUIREMENTS SPECIFICATION

Writing in 1996, Frosdick examined the proliferation of the installation of CCTV systems at football grounds and some of the problems which could result from the lack of planning in their installation and operation. He questioned whether the grounds hosting the European Championships would have any CCTV deficiencies which would affect the success of this event. As events transpired the Championships were generally regarded as a success and as an FLA Inspector I saw this at first hand at Villa Park, the first ground to install a CCTV system following the 1985 Bradford Fire Disaster. As the police commander of the ground in the late 80s and early 90s I played a part in the ongoing upgrading of the system and this process continues to the present day.

Contrast this with a visit I made to a Rugby Union Premiership ground in 2004. A full capacity crowd of nearly 9,000 was expected, but the safety officer had no CCTV system to assist in the control and management of the ground. During the event there were numerous incidents requiring stewarding action but because the safety officer was virtually blind he was incapable of making the correct decisions or using his resources to best effect. Had there been a significant incident there is no doubt that staff and spectator safety would have been seriously prejudiced. It took me back to the dark days of the 70s and 80s when policing football grounds was very much chasing shadows. All that changed with the introduction of CCTV. Sadly there are still many sports grounds that have yet to see the benefits of good CCTV and this is generally cost driven.

It remains true that there are no British Standards relating solely to the installation of CCTV at sports grounds. The best available guidance remains the 'Guidance Notes for the Procurement of CCTV for Public Safety at Football Grounds' published by the Home Office Police Scientific Development Branch. The guidance was updated in 2001 and is available to download from http://publications.psdb.gov.uk/docs/psdb09-01.pdf. Whilst written for football its content is equally relevant to other sports grounds.

CCTV installations continue to be primarily used to monitor spectator accommodation for safety purposes on the day of the event. Recording images can also be used for evidential purposes subject to the provisions of the Data Protection Acts. With many major stadia now operating as commercial enterprises, the CCTV system is used as part of the general stadium security arrangements. These are run from the control room, which is often staffed twenty-four hours a day.

Sections 16.15 to 16.19 of the 'Green Guide' answer most of the questions posed by Frosdick. The sections give clearer guidance on drawing up user requirements based on risk assessments and on how the system is to be managed and operated. A lack of proper planning, or a failure to follow the published guidance, can still result in mistakes being made with CCTV installations. Providing the 'Green

Guide' and PSDB guidance are followed there should be no real concerns about the effectiveness and efficiency of any sports ground's CCTV system. When the installation of CCTV was driven by architects and other non-users, the opportunity for mistakes has been reduced as a result of experienced users procuring the systems we see in place today.

CCTV has become an integral part of everyday life and in major thoroughfares, stores and the rail network you are never far from somebody watching you. In many ways football led the way in CCTV installations and the benefits for crowd safety and public order have become well recognised.

2.3 – PLAYING IT SAFE?

The original citation for this article is: Frosdick, S. (1997) 'Playing it Safe?', *Police Review*, Volume 105 Number 5440, 3rd October 1997, pp. 20-21.

Concerns have been raised by spectators and safety officers over the use of CS spray by officers policing football matches. Steve Frosdick reports.

Following health and safety risk assessments, a number of forces have issued CS spray as part of their officer safety programmes. All such risk management measures require particular care in the way the public are informed about them. Risk management involves the identification and management of sometimes very different perceptions about an issue – ask three different people to rate the risk of flying and you'll probably get three different answers. A holistic view is required, otherwise the countermeasures introduced to deal with one risk may themselves unwittingly create a greater risk elsewhere. The emerging debate over equipping officers who police football grounds with CS spray illustrates this.

In early July, Middlesbrough Football Club were given notice by Cleveland Constabulary that all police officers on duty inside the Riverside Stadium would be carrying CS spray from the first game of the season. The club were concerned about the implications for crowd safety.

As club safety officer, Ron Turnbull witnessed a police CS spray demonstration. 'An officer only has to withdraw the spray from his belt and large numbers of people will rapidly evacuate the area,' says Mr Turnbull. 'Just imagine the consequences if this happened at the top of a stand or concourse.'

The club did not want to see CS spray used in its stadium and sought, through the Safety Advisory Group, to enter into debate with the police. Nevertheless, the force implemented its decision.

Explaining the benefits of CS spray, Cleveland Acting Chief Superintendent John Tough says, 'Since its introduction nationwide, there's been a real marked decrease in injuries to officers. And injuries to the public are diminished compared to the use of the baton. Force policy is that officers will carry CS whenever and wherever they are on duty and that its use is a tactical decision for the individual officer to take when faces with a particular situation. When I brief my officers at Middlesbrough, I do emphasise the implications of choosing to use CS in the environment of the football stadium, where it could exacerbate the situation.'

According to officer safety expert, Inspector Peter Boatman of Northamptonshire Police: 'CS is an item of individual personal protective equipment on the same level as the baton or below. So if officers are carrying the baton on health and safety grounds, you would expect them to be carrying the CS spray if issued. CS is not a public order tactic and is not suitable for large public order situations because the spray would drift indiscriminately and contaminate the officer as well as other members of the public.'

While discussions in Cleveland are ongoing, Mr Turnbull recently brought his concerns to a Football Safety Officers' Association (FSOA) conference. The safety officers are very clear in their views. John Newsham, from Blackburn Rovers, says; 'If I was to find a couple of police officers carrying CS, I would ask them to leave. I feel strongly that CS should not be allowed in without prior notification to, and negotiation with, the safety officer. I would want consultation so that I knew that it was being carried, by whom and in what circumstances it might be used. We have professional stewarding and other control measures in place and, in my view, the risks to which police officers are exposed inside the ground are not sufficient to justify carrying the spray. Outside the ground is clearly different and is a matter for the police.'

Ged Poynton, stadium manager at Liverpool Football Club, says; 'We have an excellent working relationship with the police on Merseyside, but we are concerned about CS and have written to both the Safety Advisory Group and the police. We know the effects CS has had on other people, including children when used at events like carnivals and fetes. We believe the use of CS could cause major confusion, could result in a surge of fans and so could spark a serious safety issue. We don't want CS in our ground. Of course, we're not opposed to officers possessing nationally mandated items of equipment – we have to accept this – but we would insist on the police providing written confirmation that they accepted total responsibility for the consequences of its use.'

His Merseyside colleague at Everton is Norman Whibley, formerly lead adviser to the Safety Advisory Group in Liverpool. 'CS is the single most important safety issue facing football at the moment,' he stresses. 'In my former role, I wrote to Merseyside Police to say that I did not believe CS spray should be carried in football grounds because of the risks to public safety. And since the trials began about nine months ago, Merseyside football clubs have had an agreement with the police that officers would not bring CS into our stadia. I learned recently that pressure was coming from the Police Federation for officers to carry the spray for health and safety reasons. I have therefore written again to both the Safety Advisory Group and the police to repeat my views.'

According to FSOA chairman, Leon Blackburn; 'It seems that more and more police officers and Specials will be carrying CS in football grounds around the country. I don't want to take an obstructive view but I am anxious that there should be national guidelines drawn up so that everyone knows where they stand. And so I've written to Malcolm George [Greater Manchester assistant chief constable and ACPO football spokesman] to express our concerns and to seek an urgent meeting with ACPO.'

Mr George appreciates that 'people want to be reassured about this matter. There has been extensive training and familiarisation for officers who are fully aware of the need to consider the surroundings in which the spray may be used and of the medical treatment which may occasionally be needed in the event it is used. And there has been no incident which would support any concerns about officers carrying CS in football grounds or any other large scale crowd situations.'

Removing the sprays would cause operational difficulties for some forces. Chief Superintendent Eddie Walsh is ground commander at Blackburn Rovers. 'It may well be that ACPO could provide some national guidelines to assist individual chief officers in making their own decisions,' he says. 'But it is a very complex issue. Policing football is a major public order operation where the same officers can be deployed both outside and inside the ground. Logistically, it would be very difficult to ask officers to remove their CS spray prior to entering the ground.'

The move has also been criticised by football supporters' groups. Rogan Taylor, director of the Football Research Unit at the University of Liverpool, says he was 'dismayed about this because it sends out the wrong message about football. I can't imagine any civil disorder situation inside a ground where any sensible police officer would choose to make tactical use of this.'

Sheila Spiers of the Football Supporters Association (FSA) adds, 'Officers on Merseyside work either inside or outside the ground. Those working inside remove their sprays before going into the ground and the FSA is happy with that situation. But the FSA is not happy with the idea of officers carrying CS inside football grounds. We feel there is no way they could use it without affecting more than one person.'

Simon Inglis is a football writer and recently edited the revised version of the Green Guide to safety at sports grounds. 'Throughout the 1980s, innocent supporters were made to pay for the misdemeanours of a small minority. The use of CS would seem to be a return to the "guilty until proven innocent" syndrome. CS spray cannot discriminate between an offender and an innocent bystander. I would find its use in a British football ground quite abhorrent. At Heysel, Hillsborough and a number of other disasters, it was the people on the fringes of the main incident who died. The use of CS spray would create a potentially disastrous situation. It's not what the CS does that causes the problems, it's the people's perception of what's happening and their response to it.'

Could it be that the risks which CS sprays in football grounds pose to crowd safety are greater than the benefits it offers for officer safety?

2.3 – COMMENTARY BY JIM CHALMERS – POLICE CS SPRAY AND CROWD SAFETY

In 1997 was the article by Frosdick a 'storm in a tea cup', raising fears and concerns which really did not exist? I would suggest not and, as an Inspector with the Football Licensing Authority, I remember the discussions at various Safety Advisory Group meetings where real concern was expressed about police officers having CS spray inside football grounds. The situation was aggravated with various police forces have differing policies on whether officers would carry this part of their personal protective equipment into a sports ground.

Frosdick's 'worst nightmare' scenario of the use of CS at a football match came true in Accra, Ghana in 2000. During some spectator trouble the police fired at least twelve rounds of CS into the crowd. People sought to escape its effects and this led to a stampede during which at least 123 people died and 93 were injured, mainly as a result of crushing. More therefore than a 'storm in a tea cup'.

Fortunately in the UK his fears and concerns have not materialised as there is no evidence of police using CS spray inside a Premier or Football League ground. The only recorded incidents occurred in 1999 when at a cricket ground and a non-league football ground, some supporters turned up unexpectedly and caused trouble. When the police responded they were attacked and used CS spray in their defence. There is no report of any innocent spectators being affected by the use of the CS.

Whilst the tragedy in Ghana would be far removed from the UK police response, it is nevertheless a reminder of the effects of the dynamic energy generated by panic in a crowd situation. It is suggested that police guidelines on the use of CS spray, particularly in a confined area such as a sports ground, make it extremely unlikely that such use would ever be sanctioned in the UK. Perhaps this explains why one police commander said to me shortly after CS spray was issued to his officers, 'I won't have another Hillsborough at my ground caused by crowd panic if we use CS spray in a crowd situation.' I believe that comment would reflect the general police view and why Frosdick's concerns of 1997 are unlikely to ever materialise in UK sports grounds.

2.4 – STADIUM SAFETY: TWO CASE STUDIES IN DESIGN, MANAGEMENT AND UNDERSTANDING CROWDS

The original citation for this article is: Frosdick, S. (1995) *Stadium Safety: Two Case Studies In Design, Management And Understanding Crowds*, Report to the Building Research Establishment.

This paper presents some findings from recent research into stadium safety sponsored by the Building Research Establishment. The two case studies in this paper, derived from observations, interviews and document analysis, illustrate the need for good design to be integrated with effective management systems and an awareness of the behavioural difficulties posed by elements of the football crowd. The underlying lessons from the cases are not specific to the two grounds studied but will be of general interest to all those involved in stadium design, regulation and management.

'United' and the 'Home End'

'United' moved to their present ground during the 1930s. A sloping roof was soon added to one terrace to create the 'Home End', with space for 10,000 vociferous spectators. A riot in the 1970s resulted in one corner being caged off for the away fans. But the majority of the terrace remained the beloved home of United's noisiest supporters.

The United Chairman is a remarkable patriarchal entrepreneur. He has a clear vision for the future, including a choice of quality services for fans, the club taking real responsibility for safety and the development of the finest stadium facilities. This vision is based on both principle and a desire to expand the profitability of an already successful business. The Chairman's vision and the government's all-seater requirements sent the bulldozers into the Home End in the early 1990s.

In outward appearance, the new Home End stand resembled the 'Opposite End', creating a symmetrical appearance to the two ends of the ground. Both had goalpost roof constructions, allowing unobstructed sightlines. Both had two tiers, separated by glass fronted executive facilities. Internally however, the two stands were different.

The new stand had three zones. The lower tier, with a club members lounge and bar, was for United supporters only. Admission for ordinary matches was cheap at £6 and seating was usually unrestricted. When the ground opened, the boisterous younger fans would rush to claim the central area behind the goal. This was still the noisiest part of the ground.

The executive lounge between the two tiers spanned the full width of the stand. Reflecting the Chairman's commitment to providing a range of spectating options, £40 would buy a luxurious seat and half-time buffet.

The upper tier was specifically designed as a segregated zone for away spectators. Separate turnstiles and escalators would lead the away fans to a third floor concourse with high quality toilet and refreshment facilities. Seating and viewing quality were excellent. Halogen spotlights and seven CCTV cameras had been installed to monitor the tier. Vomitory doors were electronically controlled. At the rear was a raised patrol area for police and some of the best trained stewards in the country. In the centre of the patrol area was an enclosed control booth with built-in communication to the control room in the Opposite End.

Security consultants and police had advised against siting the away fans above the home supporters because of the risk of missile throwing. However, following his principles of customer care and the club taking responsibility, the Chairman's philosophy for the away supporters was brave and simple. The Home End upper tier was, he told me, 'the best and safest place to put the away supporters. Out of the way, so they couldn't get on the pitch, and giving them the very best of facilities.'

The police responded by insisting on tickets for the first five rows being sold only to season ticket holders of the away club and on a condition in the safety certificate allowing for the removal of seats to create a front patrol area if this was found necessary.

In terms of design for quality and safety, technology and philosophy, the Chairman's plans for the new Home End were thus at the leading edge of current thinking.

Problems and Disappointments

Having allowed home supporters to sample the new facilities for one match, the new Home End upper tier was opened for away fans. Just four weeks later, one day after a match televised from the ground, the upper tier was closed to away fans, who were relocated back to their previous position in the side stand. What had gone wrong?

It seemed that there had been problems with spectator misbehaviour. Some seats had been broken, and a few people had climbed onto the parapet, posing a risk of injury to those below. But more importantly, some away spectators in the upper tier had been spitting and throwing coins on the home fans in the lower tier.

Although the problems had not been extensive for the previous matches, the local papers had carried letters from United fans criticising the club and police for failing to intervene in such behaviour. For the televised game, the away club seemed to have ignored the condition on ticket sales for the first five rows and most of the trouble had come from those seats.

The police had felt very vulnerable with the difficult access to steep seated areas so high above the lower tier and had given up trying to get in and deal with the troublemakers. Immediately after the game, a senior police officer had begun compiling a video dossier to show to the club to seek a reversal of policy.

There was no need. The Chairman had already decided, reluctantly, that enough was enough. There had been intense supporter opposition in the first place both to the redevelopment and to having to sit down. The club could not afford the further loss of goodwill from their own fans.

With hindsight, the Chairman felt that the club had taken on too much at once. They should have concentrated on getting the home fans in the lower tier to sit down. They could then have opened the upper tier in stages, slowly building up to managing it at full capacity. The Chairman acknowledged that ease of access for police had been the least of his design considerations.

The police had not been sufficiently sold on the idea of away supporters in the upper tier. When things began to go wrong, the police, who could have called for a front patrol area, instead took the chance to say, 'we told you so'. This formed part of an ongoing political battle between police and the club. In support of his vision, the Chairman had invested substantial funds in stewards training but the police had been reluctant to further reduce their numbers and allow the stewards to take over various functions.

United supporters took the view that there had been trouble simply because the away supporters were being accommodated at the traditional Home End, where the noisiest United supporters also were. Thus the exchanges of chants and gestures was taking place between the two levels at the same end, rather than across the length of the pitch. Having the higher ground was thought to have given the away fans a sense of inviolability. It was felt that there would have been no trouble if the away fans had been in the upper tier of the Opposite End.

'City' Move to a New Ground

'City' had occupied their old ground for over 80 years. The stadium was a derelict hovel, but matched its spartan inner city environment and had an infectious and intimidating atmosphere.

'Safety Culture' at the Old Ground

The safety management system – the 'safety culture' – at the old ground was high on control but low on partnership.

The Control Room cabin was for police only. There was no stewards control. Stewards were mainly employed on passive tasks such as staffing gates and segregation cordons. The ambulance staff stayed by themselves in one corner. The fire brigade waited in a car park outside. Thus partnership between the various groups involved was low.

The police were firmly in charge of events, even perceiving themselves to have the stewards under their command. Segregation between different areas was rigid and the away fans were always kept behind at the end of the match. Control was high.

The City fans, on the other hand, had a reputation for hostile behaviour and non-compliance with authority. In practice, there was an unwritten compromise between police requirements for strong control and terrace fans opposition to it. Thus high numbers of police and rigid segregation were coupled with a laissez faire attitude to behaviour and a fire-fighting, rather than pre-planned, response to crowd behavioural difficulties. This was illustrated by police statistics which showed the highest number of police per spectator of all clubs in the region, twice the average in fact, coupled with below average ejectments and arrests inside the ground.

The New Ground

The old ground could not be rebuilt on site to comply with government requirements and supporters were resigned to moving to a new site very close by.

The new ground was described in glowing terms by one writer as 'the largest and most sophisticated football ground to have been built in Britain since 1945. ... Fully seated and covered on all four sides, with properly calculated sightlines for spectators, no restricted views and not the merest hint of a perimeter fence or crowd segregation barrier, ... the ground has four almost identical two-tiered stands, with floodlights neatly housed in the nosings of the cantilevered roofs. Front-row seats are close enough to the touchlines to remain within cursing distance'.

Toilet, refreshment and hospitality facilities were lavish. Concourses and executive lounges had a cable television system for pre-match entertainment. There was a computerised box office and the control room was filled with the latest technological systems for fire warning, intruder detection, turnstile monitoring, CCTV, public address and emergency telephones.

Thus in terms of design and technology, City's new ground, like United's Home End, was state of the art.

But buildings and technological systems do not run themselves. They need to be integrated with the safety culture.

So how had the high control, low partnership, safety culture at the old ground influenced the planning of the policing operation? How would the safety culture be affected by the radical change of environment? What impact had the old safety culture had on the design of the facilities?

Police Planning

The police had concerns that the new ground was poorly sited, with no car parking facilities for the general public, and difficult access and egress, particularly for away supporters. However, they were not intending to plan the policing of the new stadium until the close season, just three months before the first game. There was, however, a general intention to significantly reduce police numbers and hand over many functions to the stewards.

A multi-agency site visit was held about two weeks before the first match. The police spent much of the day walking around the area considering how to get the away fans between the local railway stations and the ground. The ticket office clerk at one station asked the police what they were doing. When told, she said, 'It's a bit late for all that now, love, ain't it?'

Stewards, Police and the Move to Stronger Partnership

It had been planned for the police to hand over to stewards functions such as searching spectators entering the stadium and providing an initial response to misbehaviour. Stewards would additionally be employed outside the stadium, and inside the ground on every exit gate and pitch perimeter wall gate, to deter pitch invasions. Stewards were also to be deployed at the bottom and top of every vomitory, to check tickets and enforce the prohibition on alcohol being taken into the viewing areas. Other stewards were deployed as segregation cordons between the home and away seated areas.

Police numbers had been considerably reduced in anticipation of the higher profile stewarding role. This intended change of policy would require greater police and steward interaction at operational levels and therefore suggested a shift towards stronger partnership.

The scenario then was one of strategy evolving in response to problems, a subconsciously changing safety culture and the introduction of new technology, all taking place in a brand new environment. Not surprisingly, this would give rise to operational problems.

City Stadium Monitoring Group

The creation of this group, a forum for dealing with concerns raised by local people, suggested a further shift towards stronger partnership, although the first meeting did not take place until about three weeks before the first match. Subsequent meetings, after the first two games had been played, were dominated by concerns about noise and car parking. These issues were being reactively addressed by the police and local authority.

Local Authority Safety Group

The Council intended to bring together all the various elements of the contingency plans and circulate them to all members of the Group. With three matches gone, the police still only had copies of their own and the ambulance service plans. They had received nothing else from the Council.

Subsequent meetings noted an extensive range of operational problems during the first few matches, including:

- the complete failure of the ambulance service VHF communications system;
- the general inaudibility and inadequacies of the public address system, including its complete failure at one match;
- the pitch perimeter stewards watching the game rather than deterring spectator encroachment towards the pitch;
- police horses being unable to get into the stadium because the concourse surface was too slippery for them; and
- considerable spectator confusion and the undermining of segregation arrangements, caused by a lack of information on tickets, inadequate signage and weak stewarding.

Policing and Stewarding

After various problems at early matches, the police operation was revamped and police resources considerably increased. The difficulties continued. After another match, a police debriefing noted that no stewards had been deployed to search spectators. People had been taking alcoholic drinks into the viewing areas. Concourse bars had breached licensing laws. Spectators were getting into the wrong blocks of seats. Allowing City fans into one third of the away end had led to considerable y between the supporter groups. Excited fans had been running up to the perimeter wall without steward intervention.

All the blame was being deflected onto the club and its stewards. Increasingly high control solutions were proposed. Spectators should be made to sit in designated seats. One stand should be for away fans only. The licensees should be summonsed. There should be a return to high profile policing. There was no mention of the problems on the police side.

Despite strict instructions that anyone arrested for going on the pitch should be made an example of and charged, the one fan who had been so arrested was merely ejected from the stadium. More seriously, the away fans had been kept back by police at the end of the match. Communication between the police and stewards seemed to break down, because the exit gates were still locked, creating a crushing hazard, when the fans were allowed to begin leaving.

Control Room Difficulties

The various agencies had opted for a control facility design which reflected the lack of partnership at the old ground. There were three separate rooms. The first, for the club announcer, contained the public address system and fire alarm control panel. The second, for the police, contained the police communications, turnstile monitoring and CCTV systems. The third room was divided by screens into three little 'pods' for St Johns Ambulance, the Chief Steward and Fire Brigade.

Reflecting their former position of being completely in charge, the police had sellotaped a handwritten paper sign to the entrance door to the whole complex; 'Police Control Room – Authorised Persons Only'.

Whilst the St Johns ambulance man stayed happily in his little 'pod' and the fire brigade did not attend, the increased partnership between police and stewards meant the design no longer suited them. The Chief Steward now needed to liaise with the police and use the CCTV to monitor the increased operational involvement of his staff. He now stood in the police control room, however his radio didn't work properly here.

Discussion

These cases clearly show how a lack of partnership in forward planning and risk assessment can work against the realisation of value for money from stadium redevelopments.

The engineering risk management technique of Hazard and Operability Study (HAZOPS) would seem to have a great deal to offer in these scenarios. In the case of United, a multi-disciplinary team could have sat down at the design stage with the engineering drawings, divided into suitable zones, and been facilitated through the HAZOPS risk identification process. This might be have been useful in drawing out the different risk perceptions of the club, the police and other services.

These perceptions might have varied between a team of senior personnel as against a team of front-line operational staff. The opportunities for maximising risk identification could have been further enhanced by assembling a team which was not just multi-disciplinary or mixed rank but in which different perspectives were also represented. Such a group could reflect not only the regulatory concerns of, for example, safety engineers and the police, but also the commercial interests of the club and the environmental and quality issues raised by representatives from supporter and local residents pressure groups.

In the case of City, the radical change of environment and the beginnings of a move to stronger partnership created serious operational teething problems. It seems likely that the use of HAZOPS at the early design stage may have helped to surface some of these issues and resolve them before the stadium was even built, thus minimising the difficulties experienced in practice.

The need for integrated and holistic approaches to stadium safety, which clearly emerges from this research, suggests that further work on the use of HAZOPS in the stadium environment could be very worthwhile indeed.

2.4 – COMMENTARY BY JIM CHALMERS – CASE STUDIES IN STADIUM DESIGN AND SAFETY

The case studies considered in 1995 occurred at a time when the post Hillsborough stadium reconstruction boom was at its zenith. As an Inspector with the Football Licensing Authority I saw at first hand how the scenarios in the article were re-enacted time and again as stadia were being built or redeveloped. At the time the concept of risk assessment was not prominent in the process of stadium developments and in many ways Frosdick's suggestion of Hazard and Operability Study (HAZOPS) was ahead of its time. If this process had been suggested to Club Chairmen, Directors, architects or the sports regulators they would have looked at you in wide-eyed amazement – and would still do so today. The emphasis in 1995 was on meeting the 'Taylor' requirements for all seated stadia and safe standing terraces. All of this was driven principally by cost considerations. The 1990 third edition of the 'Green Guide', with its many deficiencies, was seen as the Code of Practice to be followed in stadium design and construction and, since HAZOPS was not in the 'Green Guide', it was not a concept worth considering.

It could be argued that the 1997 fourth edition of the 'Green Guide' has embraced the principles of HAZOPS in recognising that it is only a balance of good design and good management that can achieve safety at sports grounds. Safety cannot be achieved simply by ensuring that individual components of a ground are satisfactory in themselves. The inter-relation between components is critical. None can be treated in isolation, a point well made by Frosdick in his report.

However despite the 'Green Guide' advice it is still a fact that some Chairmen, Directors, architects and stadium designers think that they know safety better than the safety professionals. Recently in a new stadium design project the architect decided where the stadium control room would be located without reference to the safety officer. In another new development the Club Chairman decided on the ground segregation arrangements without reference to his safety officer, (rather like a Chairman signing a new player without reference to the team manager). The types of issues illustrated by the case studies in 1995 are still current in 2005, so the case studies remain valid.

The saving grace in the world of football is the Football Licensing Authority who since 1991 has been and still is the catalyst for the integration of good design and management in any stadium development. Design and management are now more risk assessment-based but HAZOPS has never been embraced as a method of ensuring the proper co-ordination of design and management in projects. However that is not to say that an integrated and holistic approach to stadium design and management is not occurring as can be seen in new and redeveloped stadia the length and breadth of the country. I am sure that in the new Wembley stadium project the principles of HAZOPS are being embraced – perhaps by another name or method but with the same outcomes.

PART III - SECURITY

Introduction

Part III comprises five articles about the security of sports grounds. John de Quidt (Chief Executive of the Football Licensing Authority) said that one of the fundamental lessons to be learned from the 1989 Hillsborough Stadium Disaster was that the needs of safety and security must not get out of balance. Part III examines the history of spectator violence and how the control of alcohol can impact on spectator behaviour in sports grounds. There is a detailed explanation of a club-orientated ticketing and marketing system (COTASS) introduced in Holland and a comparison of both the UK and American responses to stadium security in the aftermath of 11 September 2001.

3.1 – Review of Spectator Violence

In 2002 Frosdick wrote an excellent summary of the causes and control of spectator violence. He outlines the difficulties in defining the problem and gives an overview of the various academic explanations offered for the phenomenon, which he shows is global in nature. The debate on spectator violence continues today and the article provides a succinct review of the problem in a historical context.

3.2 – Alcohol in Stadia

In 1998 Frosdick argued for a relaxation of the licensing laws relating to the sale and consumption of alcohol in sports grounds. Seven years later in 2005, new legislation provides for 24 hour drinking with licensing regulation passing to local authority control. Comment is made on whether this relaxation of the licensing laws should also be applied to sports grounds.

3.3 – The Dutch COTASS Project

This 1997 article reports on the COTASS project pioneered by the Royal Dutch Football Association. Frosdick examines the concepts of club cards, national and local ticketing, access control and relationship marketing. The system had a mixed reception in Holland and, although some UK clubs have introduced club cards, the COTASS concept has not been taken up elsewhere.

3.4 and 3.5 – Venue Security After 11 September 2001

These two articles from 2002 reflect venue security reactions to the September 11 atrocity in New York. Views expressed at Conferences in London and Arizona are examined and reflect the differing perspectives on increased venue security between the two countries. Comment is made that in the UK venue management do not appear to be doing enough to assess the terrorist threat and to plan for such a contingency.

3.1 – VIOLENT BEHAVIOUR

The original citation for this article is: Frosdick, S. (2002) 'Violent Behaviour', *Stadium & Arena Management*, October 2002, pp. 25-27.

Steve Frosdick gave a paper on the causes and control of spectator violence at the 'Stadia & Arena 2002' conference in Portugal.

Recent events show that spectator violence remains a key concern for stadium and arena managers. In order to better deal with the problem, it is important to understand why spectators sometimes behave as they do.

This paper will first set out the current and historical nature of sports-related violence and highlight some of the difficulties in defining the problem. Second, it will review the various theories which have been offered to explain the problem. Third, it will go on to refer to the methods used to try and control the problem. This is an ambitious agenda for a short paper, so we will be taking a fairly brisk gallop through the issues.

Spectator violence at sports events is a current problem – in England, across Europe and indeed throughout the world.

In 2002, an English FA Cup-tie between Cardiff and Leeds was disrupted by persistent missile throwing and concluded with a large-scale pitch incursion. And serious fighting outside the ground marred the end of season play-off between Millwall and Birmingham in May.

Examples from 2002 in Europe come from Scotland, Holland and Spain. Aberdeen versus Rangers in January saw missile-throwing and fighting inside the ground, whilst an Ayr United match in April was abandoned after a protest-related pitch invasion by Airdrie supporters. The UEFA Cup Final between Feyenoord and Dortmund involved fighting outside the ground and in the city of Rotterdam. And Spain's end of season saw fighting inside and outside grounds, together with attacks on players and officials.

Following the previous suspension of the Argentinean League because of crowd violence, renewed problems in the 2001/2002 season included the abandonment of a derby match between Boca Juniors and River Plate.

But the problems are not confined to soccer. American football saw serious disorder at a Cleveland Browns match in December 2001, when sustained missile throwing forced the players to leave the field. Even at the Winter Olympics in Salt Lake City in February 2002, the police fired tear gas to disperse disorderly crowds after the finals of the bobsleigh competition!

Spectator violence is also an ancient and historical problem, going back to when there was disorder between the 'Blues' and the 'Greens'; the supporters of different chariot-racing teams in Ancient Rome.

Many modern ball games are derived from the medieval folk-football played in England since the 13th century. This was an excuse for fighting which regularly featured violence, death, injury and damage. There were frequent attempts to ban and suppress it. In 1314 the Mayor of London issued the following proclamation:

'And whereas there is great uproar in the City through certain tumults arising from the striking a great footballs in the field of the public – from which many evils perchance may arise – which may God forbid – we do command and do forbid, on the King's behalf, upon pain of imprisonment, that such games shall not be practised henceforth within this city.'

So right up through history, we find that football – and the other ball games to which it gave rise – have been associated with violence. The worst incident I've heard of came from the island of Mauritius on 23 May 1999. Following a soccer match between two teams named 'Scouts Club' and 'Fire Brigade', there were riots in the capital, Port Louis, during which seven people were burned to death.

Notwithstanding this long history, there are real difficulties in defining the problem. In the first place, what do we call it? The particular association with football caused the media to invent the label 'football hooliganism', but perhaps 'spectator violence' would be a more accurate name?

The second problem of definition is that there are so many variables. Are we talking about offences against the person (e.g. assaults), against property (e.g. vandalism), or against the State (e.g. disorder)? Do we limit the definition to acts which are crimes, or do we include misdemeanours or even plain anti-social behaviour? And where do the behaviours take place – inside, outside or even away from the venue? Are we only counting organised violence, or do we include the spontaneous, e.g. a pitch incursion to celebrate a goal? Do we include 'contagious' acts where crowds 'go mad' because they catch the bad behaviour off each other?

So we're not sure what to call it, we're not clear what 'it' is, and to make matters worse, we find that the statistics about it are wholly unreliable!

There is a problem with what gets counted. The English police include 'ticket touting', 'forgery' and other non-violent offences in their official statistics for football-related arrests. And they don't keep records for other sports.

There's also a problem with what criminologists call the 'attrition rate'. Not all the offences committed get reported to the police, who are then selective about what they record. Of those recorded only a small percentage get cleared-up. It's said that, for every 100 offences committed in the United Kingdom, only 2% get cleared up. So a lot of offences of spectator violence never make it into the figures.

The statistics are also more a reflection of police tactics than of the extent of spectator violence. The Head of the UK National Criminal Intelligence Service commented that sometimes the police will make large numbers of arrests to intervene and prevent disorder, whilst at other times they will disperse a disorderly crowd and make few or no arrests.

Explanations as to who the hooligans are and why they do it are equally problematic. There are simplistic populist explanations and there are more complex academic theories.

One populist explanation is that 'all sports fans are animals'. They behave and so should be treated as such. Yet research clearly shows that many sports fans are women and that all ages and social classes attend sporting events.

Another populist theory blames alcohol for violence. Alcohol is often banned even though this flies in the face of the research evidence. Alcohol may be a contributing factor at football in some Anglo-Saxon cultures, but the Danes, Dutch and Irish manage to drink vast amounts without particular problems of violence. And sports such as rugby have strong links with alcohol but with very few problems of violence.

A populist police explanation claims that sophisticated organised gangs orchestrate the violence. Well the police would say that, wouldn't they, since it justifies the whole industry they've created to deal with soccer hooligans. In the UK there are good examples of intelligence between club safety officers being rather at odds with the police intelligence for particular matches.

English academic explanations have been marked by the acrimonious and vitriolic nature of the debate both in the literature and at conferences. Rather mirroring the object of their study, English academics have focussed as much on putting the boot into each other's theories using obtuse sociological jargon as they have on providing intelligible explanations.

In 1971, Professor Ian Taylor took a Marxist sociological perspective, suggesting that violence was 'lumpenproletariat' protest against the hijacking of football by big business. In 1978, John Clarke's sub-cultural perspective concluded that football-related violence was associated with the skinhead style. Both of these explanations were criticised as politically motivated theories with no real supporting evidence.

Also in 1978, Peter Marsh and his social-psychological colleagues constructed the idea of 'aggro'. They observed Oxford United fans and concluded that what was misconstrued as violence was actually an illusion involving ritual displays of aggression between young men. During the 1990s, the social anthropologist Gary Armstrong followed Sheffield United and concluded that what was going on was a disorganised acting out of working class male identity rituals. Both Marsh and Armstrong were criticised for trying to over-generalise from studies at just one club.

The most prolific and perhaps plausible English academics were Eric Dunning, Patrick Murphy and John Williams from the University of Leicester, Writing in the 1980s and early 1990s, their social historical perspective focussed on social exclusion and the cultural traditions of the 'uncivilised' rough working class: maleness, solidarity and aggression.

In 1992, the journalist Bill Buford took a more populist 'biological' perspective, describing the adrenalin rush brought on when it 'goes off'. More academically, the psychologist John Kerr wrote about the hooligans' search for 'peak' & 'flow' experiences, but his reasoning was criticised as *a priori*, i.e. there was no evidence for his theory except itself.

There is also convincing evidence of media amplification leading to moral panic and the self-fulfilling prophesy of violence. Media reporting suggests the likelihood of disorder. Football therefore becomes more attractive to the type of person disposed to violence. The police plan for the trouble anticipated and may be inclined to over-react to minor incidents. Nobody is surprised when serious disorder breaks out, since it was what everybody expected in the first place. And the press – who started it all – feign outrage at 'England's Shame'.

Despite academic denials, there are commonalities in all this and it is fairly straightforward to merge these various English theories together. Taking an outrageously reductionist approach (for which my academic colleagues would rightly slay me), we might conclude that violent English fans are male, young, and working class; that they are acting out something to do with post-modern masculine identity; and that they do it because it's fun.

Italian academic research has studied the *tifosi* and the *ultras*. There are two matches going on – one on the pitch and one involving the fans, who act out friend/enemy rituals against their historical rivals. Other European research has been more limited and there is insufficient space to go into it here. The key point is that there is no single Europe-wide theory and that the explanations of violence rely on different factors in different countries. An excellent 1996 summary by Carnibella *et al.* concludes that

'It is clear that no Europe-wide explanatory framework has yet been developed. It may be the case, given the distinctive nature of ultras, hools, roligans, etc. that such a framework may be unachievable or inappropriate. The social and psychological factors which lie at the root of football violence in, say, Italy, may be quite different from those which obtain in Germany or Holland. The football stadium provides a very convenient arena for all kinds of collective behaviour. There is no reason to suppose, therefore, that the young men who use such arenas in different countries are all playing the same game'.

One significant recent contribution to the debate comes from Professor Eric Dunning in his 2000 article 'Towards a Sociological Understanding of Football Hooliganism as a World Phenomenon'.

'It is important to stress that it is unlikely that the phenomenon of football hooliganism will be found always and everywhere to stem from identical social roots. As a basis for further, cross-national research, it is reasonable to hypothesise that that problem is fuelled and contoured by, among other things, what one might call the major 'fault lines' of particular countries.'

Dunning illustrates this 'fault-lines' hypothesis with reference to social class and regional inequalities in England and to religious sectarianism in Scotland. Elsewhere in Europe he mentions linguistic sub-nationalisms in Spain and the Italian rivalries between cities and between the North and the South. Similar tensions are evident between the former East and West in Germany and between the political left and right.

Extending this thinking, we might consider race relations and the resurgence of the far right as a major 'fault-line' in France, recalling that the most serious recent incident there was the abandonment of the France v Algeria match at the Stade de France.

Further afield, religious communalism is the major flaw in Mauritian society, where terrace violence involving supporters of communal teams – Hindu, Moslem and Creole – increased during the 1990s in parallel with increased communal identities.

Turning now to the question of social policy and control, we find that most social policy has been negative and focussed on repression – as it was with folk football in the Middle Ages.

In the UK and United States, we find assertive policing and stewarding. Exuberant behaviours are tolerated up to a point. But once the threshold of tolerance has been breached, the culprits experience a swift 'no nonsense' ejection from the venue.

There are examples of serious police over-reaction, e.g. in Belgium during the 2000 European soccer Championships. In one notorious incident, Belgian police threw tear gas grenades into a crowded bar and indiscriminately arrested everyone inside.

There was some nervousness about the policing for the 2002 World Cup in Japan and South Korea. Early in the preparations, senior Japanese police officers were asked, 'What level of tolerance will you show to supporters?' They simply did not understand the question. Fortunately, their understanding moved on and, as it turned out, the fans behaved very well.

The UK has seen quite draconian legislation involving travel restrictions, not only on convicted hooligans but also on unconvicted fans whom the police suspect may cause trouble. Because of moral panic, the criticisms of this infringement of people's civil liberties have been rather muted.

We've already spoken of Argentina, but in Mauritius, following the Port Louis deaths, all organised football was suspended May 1999 to November 2000 – a whole season – while the communal teams were reconstituted on a geographic basis. This transformation resulted in a massive loss of revenue and fan interest, but the evidence is that violence has been eliminated. Longer term, it is hoped that the fans will return.

There have been other good examples of more positive approaches to social policy.

The Dutch arrangements for Euro 2000 contrasted markedly with the Belgian. In Holland, careful planning and a friendly but firm approach resulted in no reports of serious problems.

Control of violence has also become a sub-set of more general risk and safety management. This is particularly so in the UK where there has been a post-Hillsborough paradigm shift from 'protection from the crowd' to 'protection of the crowd'.

There is also evidence of efforts to change the culture of football supporting. Scotland's Tartan Army reinvented themselves as the 'friendly' fans in order to beat the English off the pitch. There are strong initiatives among England supporters to change the culture of following the national team abroad – cultural tourists rather than invading hordes. And there are many examples of fan coaching projects in Europe. These involve clubs and municipalities employing social workers to liaise with fan groups to facilitate atmosphere but discourage violence.

All in all then, we have seen that there are problems defining the nature of spectator violence, that there are multiple and complex explanations of the causes of the violence, and that there are both repressive and more positive ways of dealing with what is both an ancient and yet current world problem.

Further Reading

For a more detailed analysis of football-related violence, see Frosdick, S. and Marsh, P. (2005) *Football Hooliganism*. Cullompton: Willan Publishing (ISBN 1-84392-129-4).

3.1 – COMMENTARY BY JIM CHALMERS – REVIEW OF SPECTATOR VIOLENCE

In this 2002 article Frosdick achieves the impossible by providing an excellent summary on the causes and control of spectator violence, which has been with us since the time of chariot racing in ancient Rome. It is a topic which over generations has been the subject of numerous inquiries into football-related violence and disorder stemming from the 13th century and more recently in the 1960s when the media penned the phenomenon as 'football hooliganism'.

Frosdick's article summarises the academic arguments on the causes of the phenomenon from a sociological, social-psychological, biological, anthropological and criminological perspective – but with no one explanation providing a full insight into the problem which persists in football today. The issue becomes even more confused with no clear definition of what is meant by 'spectator violence' or 'football hooliganism'. Frosdick examines the problem in an international context with examples of spectator violence far more acute than the UK experience.

The complexity of the subject can be seen from the disagreement amongst academics on the cause of the problem and conflicting views on the need for any further research into the phenomenon. One group of academics argue that the subject is probably over-researched with little in the way of new insights now forthcoming. Another group take a differing view arguing that since spectator violence is a world-wide phenomenon, a view shared by Frosdick, then research would benefit from a world-wide examination of the problem. Instead of the vitriolic debate between academics, which Frosdick describes, it could be argued there is a need for a holistic, global research programme if any new insights or solutions are to be found relating to spectator violence.

Frosdick explains that throughout history and up to the present day the problem has been and still remains that 'spectator violence' has defied definition. I would argue that unless the problem is defined then measurement and explanation of causation will remain difficult, since in 2005 no one is any the wiser as to what 'spectator violence' or 'football hooliganism' really means. In a recent research paper I argued that 'football spectator violence' could be defined as, 'All assaults or threats involving members of the public which occur at or whilst travelling to or from a regulated football match, of which the police are aware and which may require police intervention or action in partnership with others'.

This definition may not be perfect but it is an attempt to define a problem, which up to now appears incapable of definition either by the sport regulators, police, government or academics. The problem becomes even more acute, as Frosdick rightly says, in any examination of football-related arrest statistics, when violent disorder can mean anything from a shove to a serious assault.

Frosdick's article does not offer a solution to spectator-related violence nor would it be expected to. What it does do is to bring into focus the many diverse opinions which exist and the extent of the problem in 2002. The content is equally valid in 2005. The article clearly illustrates that, as society changes, so will the phenomenon of spectator-related violence. As someone involved in policing, regulating and managing safety at football matches between 1960 and 2005, I know at first hand how spectator behaviour has changed in those 45 years, and I would suggest will continue to change in the years ahead. Frosdick provides an excellent historical review of spectator-related violence, but it is unlikely that a solution will be found without further research.

3.2 – DRINK OR DRY?

The original citation for this article is: Frosdick, S. (1998), 'Drink or Dry?', *Stadium and Arena Management*, Volume 2 Number 4, August 1998, pp. 20-24.

At the recent European Stadium Managers Association Convention in Paris, Steve Frosdick argued in favour of relaxing alcohol controls.

Restricting or even prohibiting the possession and consumption of alcohol in stadia is one of a number of management controls intended to counter the risks associated with crowd violence and misbehaviour.

Of course spectator violence is nothing new. Disorder has been associated with sport since before the supporters of the 'blues' and the 'greens' fought each other at chariot races in ancient Rome. And crowd violence has occurred at sports such as rugby league, cricket and boxing in England, rugby in France and baseball and American football in the United States. Football-related disorder has also affected virtually every country in which the game has been played.

The debate about spectator violence and alcohol has been reflected in many official reports in England and Scotland over the last thirty years. The question was also addressed in the 1985 European Convention on Spectator Violence. The various arguments and recommendations contain common themes: that alcohol consumption is an important factor in crowd misbehaviour, particularly at football matches; that controls of some sort need to be introduced; and that these should include both restrictions on sales and the creation of specific criminal offences.

In Scotland there has been an absolute ban on alcohol in stadia since 1980, and this applies to both rugby and football. In England, the law allows Licensing Magistrates to grant an exemption from the ban and thus allow sales under certain conditions. It also creates a number of criminal offences, for example to be drunk inside the ground; or to possess alcohol in any part of the ground that offers sight of the pitch.

Where Are We Now?

Today, we find a situation where we have various approaches across different countries, between different sports in the same country and even within the same sport or indeed within the same stadium, depending on the type of match being played. Whilst some countries have a complete ban, others have different laws in different places and for different sports. In the United States laws restricting alcohol vary from State to State and from sport to sport. There are even major differences within the same State. In Texas, for example, the local ordinances vary dramatically, even between political precincts within the same town. In Dallas, some parts of the city are completely 'dry', others allow only the sales of wine and beer and yet others have no restrictions at all. At the Texas Stadium, home of the

Dallas Cowboys, alcohol was banned until about three years ago. Now beer can be sold, but not 'hard liquor'. So some stadia have no alcohol at all, whilst in others you can even order beer to be served to you in your seat!

At football matches played under the jurisdiction of FIFA, such as the 1998 World Cup in France, the sale and possession of alcohol in the stadium is prohibited by the rules of competition. Football matches in the UEFA competitions are covered by UEFA regulations on safety and security in the stadium, which provide that 'No public sale or distribution of alcohol shall be permitted within the stadium or its private environs.'

So when Wembley Stadium, for example, is staging a FIFA or UEFA match, no alcohol is allowed. Yet at the FA Cup Final, alcohol is sold at the concourse bars, although the law does not allow spectators to take their beer back to their seats, since they would then be within sight of the pitch. However, when rugby matches are played at Wembley, this restriction does not apply and so rugby fans could in theory drink their beer whilst watching the game. The same anomaly is found at several English stadia where both rugby and football are played.

Watford football club is also home to Saracens rugby club. Here the stadium manager is happy with the differences in the law and would not like to see it changed to let football supporters take alcohol back to their seats. He says that they have different types of supporters for rugby and football and that they never have any trouble with rugby fans.

Loftus Road is home to both Queens Park Rangers football club and Wasps rugby club. Here the stadium manager originally stewarded the rugby crowd the same as the football crowd, but found this was quite unnecessary. They now use considerably fewer stewards and have no segregation at rugby games. At one recent match, the rugby crowd was about one third the size of the previous day's football crowd, yet they bought twice as much alcohol and caused no problems at all. The stadium manager would not like to relax the controls at football matches, as he feels this would bring back the previous problems of football crowd misbehaviour.

So how effective are the current controls? Referring to the complete ban in Scotland, the report into the 1989 Hillsborough disaster said that there was no doubt that 'this measure has greatly reduced the problem of misbehaviour at Scottish football grounds'.

There is an argument, however, that such blanket bans are too simplistic. At the Olympia Park in Munich, the stadium management and police used to sit down before each game and assess the risk of disorder before determining whether alcohol sales would be allowed. So Bayern Munich v FC Kaiserslautern would be fine for alcohol sales, but Bayern Munich v 1860 Munich would not. This seems to me to have been an entirely sensible risk-based approach. However, 18 months ago, a new police chief arrived and directed that no alcohol at all be sold either in the stadium or in the surrounding Olympia Park for any football match. Fans are free, however, to congregate in the bars just beyond the park where they can and do drink themselves silly.

Stoke City football club, who have just completed the first season in their new ground, provide an interesting case. At their old ground in the town centre, all the catering outlets had at least a small view over the pitch and so it was not possible for the club to get permission to sell any alcohol in the ground. According to the stadium manager, many of the fans remained in the local pubs until 2.55 pm, arriving very late and causing long queues at the turnstiles. The police often asked for the kick-off to be delayed because large numbers of fans were still queuing to get in. On one occasion, a public address announcement was made at 2.55 pm that the kick-off would be delayed until 3.15 pm. The fans waiting outside promptly turned round and went back to the pubs until 3.10 pm!

The new Britannia Stadium is built on the site of an old coal mine about 20 minutes walk from the town. Parking is difficult and the only public transport is a bus service. Having a new ground meant the club were now able to get a liquor licence. Alcohol is sold up until kick-off, from 15 minutes before until 15 minutes after half-time and then again after the match. The fans now tend to arrive early to enjoy a drink before the game, at half-time and even after the match, although this tends to be only when their team have won. There have been no arrests for drunkenness and no matches when the sale of alcohol is prohibited. Sales were even permitted for the last game of the season, against Manchester City, when very serious disorder was anticipated and indeed did break out. In fact more than 20 people were hurt, 300 ejected from the ground and 15 arrested. But the stadium management felt that a ban on alcohol at the ground would have caused very serious problems in the town. It was better to get the fans in and control them at the stadium.

Relaxing The Controls?

As I have argued in previous articles, stadium managers are required to strike an appropriate balance between four competing demands. Commercial pressures require them to optimise the commercial viability of the venue and its events. Spectator demands for excitement and enjoyment require credible events staged in comfortable surroundings. Regulatory and other requirements for safety and security must also be met, whilst any negative effects which the venue and event may have on the outside world must be kept to a minimum. The key to striking a balance lies in taking a holistic approach to risk assessment and risk management, ensuring that management controls in one area do not have unnecessary adverse consequences in another.

Stadium managers faced with an event where the risk of spectator misbehaviour is high, have a whole basket of management controls to choose from. These cover such areas as ticketing, signage, access control, deployment of police and stewards, spectator segregation and protection of the playing area, as well as alcohol controls. I believe that stadium managers and the police should consider each event on its merits, selecting the control measures which best suit what is known about the location and layout of the stadium, the nature of the crowd attending and the event itself.

Alcohol controls which the police and stadium management feel are not needed for that event therefore in my view create unacceptable risks elsewhere. For the commercial manager, a ban on alcohol sales creates a risk to revenue. The fans will not buy cola instead. They will drink outside and come in late, reducing what I believe is known in the trade as 'ancillary spend per head'. For many spectators, a ban on alcohol quite simply reduces their enjoyment of the event, and where it has been needlessly imposed from outside, may create feelings of resentment. And local residents and businesses have to put up with more noise and disorder around the ground than might otherwise be the case. From a safety point of view, there are compelling arguments that a total ban results in late arrivals and a last minute rush to get in at the turnstiles.

But even restrictions on when alcohol can be sold and on where it can be consumed can themselves cause safety risks. According to John Beattie, the stadium manager at Arsenal FC, the rush to buy and consume alcohol at half-time causes large crowds and near crushing on the concourse, which is often blocked by people drinking alcohol. This also causes difficulties for other people wishing to purchase food or use the toilets.

Disagreeing with their colleagues at Watford and Loftus Road, the stadium managers at both Arsenal and Stoke City football clubs would like to see the law changed so as to allow football fans to take their beer back to their seats. They see no good reason why football fans should be treated differently to rugby, believing that all fans should be treated with respect and provided with decent facilities.

The key to relaxing the controls, they believe, lies with effective stewarding. Stoke City have stewards whose only job is to prevent people taking beer back to their seats. Such stewards could surely be trained for the job of preventing drunken fans from entering the stadium in the first place - which must be the most sensible control of all - and for monitoring fans for drunkenness in the concourses and controlling what was being taken into the viewing accommodation.

Nottingham Forest Football club have for some years also had specially trained 'alcohol stewards' deployed in the concourse areas. The safety officers at Nottingham Forest have developed a particular expertise in helping other clubs to gain permission to sell alcohol in their stadia. They have a ten-point plan of licensing conditions, namely that:

- Alcohol bars should be separate from fast food outlets;
- Notices should be displayed showing the serving times and the maximum number of purchases by one person;
- Only beer, lager and miniature plastic bottles of spirits should be sold;
- Each person should be allowed to buy no more than four drinks at a time;
- Beer and lager should be served in disposable plastic glasses (and it is impossible to carry more than four of these at a time);
- All bars should close five minutes after the start of the second half, with twenty minutes then allowed for customers to finish their drinks;
- Alcohol consumption should be strictly confined to the permitted consumption area;
- There should be adequate signage to define the consumption area;
- There should be an adequate number of clearly defined stewards to supervise each area; and
- Each bar should be monitored by CCTV cameras.

Whilst not a universal panacea, these ten points do offer a helpful framework within which stadium managers can consider the (re)introduction of alcohol sales.

Alcohol controls may well be appropriate for some events. But for many others, they may not. Blanket bans and prescriptions on when and where alcohol may be sold are thus in my view not helpful. What is needed instead is the sensible assessment of risk, together with legislative frameworks which allow managers to select the control measures appropriate to the individual stadium and event.

3.2 – COMMENTARY BY JIM CHALMERS – ALCOHOL IN STADIA

In this 1998 article Frosdick examines the sale, possession and consumption of alcohol in the stadium and its relationship to crowd violence and misbehaviour. He examines how levels of alcohol controls relate to the management of crowd behaviour risks. The article stems from a paper at an international conference in Paris and this perhaps explains the references to the French and American experiences of alcohol in stadia. As someone who has attended sporting events in America, Germany, France, Holland, Italy, Guatemala and Cyprus both as a police officer and as a sports regulator, I am not certain that comparing the experiences of alcohol in sports grounds in other countries necessarily has any bearing on UK experiences or can provide any lessons for us. Crowd behaviour is different as are attitudes to alcohol – a debate which also relates to the drunken behaviour of many UK holidaymakers abroad. In this regard it is perhaps surprising that Frosdick did not explain in more detail the outcome of Chapter 6 of the Taylor report into the Hillsborough stadium disaster, which details the history of the control and regulation of alcohol in sports grounds based on crowd behaviour. However that is in the past and the article examines proposals whereby good management practices in the sale and supply of alcohol can be of benefit to both sports grounds and their customers.

It is interesting that this article in 1998 in favour of relaxing alcohol controls in the stadium may have a parallel in 2005 with new UK legislation which will allow 24 hour drinking in licensed premises. The legislation also changes the liquor licensing body from the judiciary to local authorities. I think it would be a step too far to see sports grounds licensed to sell alcohol all day and night but the change in the licensing body could benefit sports grounds. Since 1975 the local authorities have had the responsibility for the safety certification of sports grounds. In my twelve years as an Inspector with the Football Licensing Authority there were numerous examples of conflict between spectator safety requirements and the provision of alcohol at sports grounds because of the conflicting requirements by two sets of regulatory bodies. It therefore makes more sense for the local authority to undertake both regulatory functions.

UK Government views on the relaxation of licensing hours could support Frosdick's argument for greater deregulation of the sports grounds control of alcohol legislation. Sports grounds are after all intended to be places of public entertainment and being able to have a drink is part and parcel of the public entertainment scene. Why should football fans therefore be treated differently from fans watching a cricket or a rugby match? In particular anyone who visits football grounds on a regular basis will have some empathy for the comments in the article about overcrowding and crushing in stadium concourses in fifteen minutes at half time, when the desire of hundreds of fans to get a drink is aggravated by the fact they cannot then take it back to their seat – unlike in rugby. Giving the local authority both the liquor licensing and the spectator safety functions will cause this risk to be examined more closely.

Whilst police football arrest statistics are notoriously a poor barometer upon which to judge spectator behaviour, nevertheless in the 2003/04-football season out of a total attendance of 29,197,510 there were only 599 drink related arrests inside football stadia. I am sure that compared to the arrests in our City centres for drink-related offences this number pales into insignificance. Since football is the most regulated sport in the UK with all Premier League and Football League grounds covered by CCTV and effective stewarding it could be argued that the sale and supply of alcohol inside stadium is also over regulated. The time may well now be right for Frosdick's argument for a more relaxed approach to be adopted.

Figure 3.2.1 Drinking alcohol in sight of the playing area has not previously been allowed in UK soccer stadiums.

3.3 – ROYAL DUTCH: INTEGRATED TICKET SYSTEM

The original citation for this article is: Frosdick, S. (1997) 'Royal Dutch: Integrated Ticket System', *Stadium and Arena Management*, Volume 1 Issue 6, October 1997, pp. 32-36.

Ticketing, access control and knowing your customers are all important in public assembly facilities management, but there are often tensions between safety and marketing. Through the COTASS project the Royal Dutch Football Association are pioneering an integrated approach to these areas.

Much of my work in the stadium and arena industry has focused on the balancing act which PAFs (public assembly facilities) managers have to strike between commercial viability, customer enjoyment, safety/security and the local environment. So it was particularly interesting to meet Folkert Schroten from the Royal Dutch Football Association (KNVB) at the recent European Stadium Managers Association (ESMA) Conference in Barcelona. Folkert is project manager of the KNVB COTASS project, which is using database and networking technology to encourage Dutch football clubs to introduce an integrated approach to three of these areas, through access control, customer service, ticketing and marketing.

COTASS – 'Club-Oriented Ticketing and Authorisation System for Stadia' – has its origins in 1992, as an initiative started by five clubs concerned about spectator misbehaviour. A previous attempt to introduce an access control card had been scuppered on day one by concerted supporter action. But there was (and in official quarters still is) a view that requiring all supporters to possess some kind of card to gain access to a football ground would act as a curb to soccer violence. This is a questionable assumption, since much of the fighting in Holland happens away from the ground and many fans complain that it is unfair to restrict the liberties of the many because of the sins of the few.

Nevertheless, COTASS was set up within the KNVB with the primary objectives of increasing security and customer service whilst at the same time increasing commercial opportunity through ticketing and marketing. From the outset, it was appreciated that for COTASS to address its security objective, it had to deliver commercial benefits to clubs and service benefits to fans. According to Folkert Schroten, 'If you ratchet up security, you have to ratchet up service at the same time to keep the support of the fans.'

Although the project has been running for nearly five years, it is still relatively undeveloped. Holland is a liberal democracy with a tradition of relaxed tolerance rather than rigid enforcement of rules. So the team have deliberately operated with a light touch, facilitating progress in small steps rather than trying to impose the whole system on everyone at once. Progress has thus inevitably been mixed. Let's take a look at the theory behind each of the five areas and assess where things appear to be in practice.

Club Card Credit

To operate the Club Card scheme, clubs must purchase the COTASS software, which costs Dfl 50,000 ($25,000 or £15,500) for a four year licence and the support of the COTASS team. The software is a relational database held on an ordinary personal computer, with communications provided via a modem and X400 e-mail system.

Club Cards were introduced to all clubs in the Dutch Premier Division in 1996/97 and are being introduced this season throughout the First Division. The card costs Dfl 12.50 (about $6 or £4) and is valid for two seasons. There are different types of Club Card: the normal card, supporters club card, season ticket card and business card. Cards are standard credit card size with a magnetic strip and clubs can use whatever different designs they wish. Some clubs are piloting the use of photographs as an additional card security measure.

Application is made on a national form and each person is only allowed to hold one Club Card for one football club, although there is a separate Orange Card for supporters of the Dutch national team. In the interests of protecting personal information, each club database holds only the details of its own cardholders. The central COTASS database holds only enough personal details to allow for checking for duplicate cards, or where a card has been withdrawn for security reasons, to prevent blacklisted applicants from obtaining another card. Cards are produced by a central supplier and then either mailed out direct or else, in the case of season ticket cards, distributed through the clubs.

Whilst over 400,000 cards have been sold, a recent TV report showed that there is some resistance from the fans to the Club Card system. The COTASS team hope that this is a temporary trend, and that fans will eventually see the advantages of choosing to hold a card.

National Ticketing

This is the most fully implemented and successful part of COTASS. For each match each club 'defines the event' in its COTASS database to show who will be allowed to buy tickets and when and where they will be allowed to buy them. These details are then uploaded onto the national lottery network and tickets for the match are then sold through each of the 900 lottery outlets across the country.

Let's say that Feyenoord of Rotterdam are playing PSV Eindhoven in three weeks time. Feyenoord could decide that in two of their stands, only Feyenoord card holders can purchase tickets and that for the first week only Club Card holders will have the right to buy. They could allow holders of any card except PSV to buy tickets for the third stand, and restrict sales in the fourth stand to PSV card holders who at the same time buy a voucher for the official PSV away travel package.

It's just like buying a theatre ticket. Every lottery outlet has a plan of every stadium. Your card gets swiped into the lottery terminal, and the network checks your card is not on the whitelist of lost/stolen cards or blacklist of banned cards. You choose the match you want to buy for, your ticketing options come up on screen, you choose where you want to sit and your ticket is printed. This is a real customer service benefit for fans. Within the restrictions set by the clubs themselves, every fan can buy a ticket for any match from anywhere in the country. Every club now participates, the system works well and the fans like it.

There are safety benefits too, since very ticket has a unique barcode and the national network maintains a record of which card purchased which tickets. In the event of an incident, the club will know that, for example, AFC Ajax card number 123456 bought the tickets for those seats. Whilst this does not tell you exactly who is in the stadium, it does provide an audit trail to assist an investigation.

Local Ticketing

Having a Club Card is of course no guarantee of a ticket. In the long term, the intention is that it will not be possible to buy a ticket for a professional football match in Holland unless you hold or have applied for a Club Card. In the short term, it will still be possible for non-card holders to buy tickets at the stadium on the day of the match although at some clubs this is only allowed if the supporter pre-pays the Club Card fee in return for a voucher. If they choose not to apply for a Club Card, they have to pay the extra Dfl 12.50 every time they go to a match.

This is not to say that only Club Card holders can go to football matches, since for many games, holders can buy more than one ticket. And some clubs still sell cash tickets without requiring pre-payment of the Club Card fee. So they system deliberately has many loopholes since, according to Folkert Schroten, 'the social costs of 100% control are simply not acceptable'.

But continued problems with hooligan behaviour, particularly as a result of a fight between Ajax and Feyenoord fans in which one person was killed, are increasing the pressure to ratchet up security. I visited the first match of the 1997 season at NEC in Nijmegen where, for the first time, portable scanners on a serial network were being piloted to check that people buying tickets at the booths held valid Club Cards. Inexplicably, only one booth was left open for non-cardholders, who at first had to pay the Dfl 12.50 supplement. However the resulting queue was so bad that the experiment had to be abandoned for safety reasons and the kick-off delayed for ten minutes.

Access Control

Access control comes in two stages; ticket purchase and physical access to the stadium. The theory is that ticket purchase is controlled through the Club Card scheme and that access to the stadium is controlled by stewards equipped with barcode scanners. Before every match, details of the latest white and blacklists are downloaded from the national network onto the local club database. Each scanner is linked to the database to check that the tickets presented are still valid.

But the costs of the hardware and of hardwiring the stadium from the club database to each access control point are considerable. Clubs are supposed to be installing the system, but the COTASS team have found them very reluctant to do so. So far, only three Premier Division clubs – Ajax, Feyenoord and Volendam – have invested in the technology and are using it, and only one club – ADO Den Haag from the First Division – are currently installing it. A few clubs have implemented small-scale system, for example to control access to business club lounges.

There are very few turnstiles installed at Dutch football grounds. Typically, supporters seeking access to the stadium have to present their ticket for checking then, quite often, be rubbed down for weapons, etc. by a steward. Where it is installed the scanning system does not seem to have caused any increase in queuing simply because it takes longer to do the security check than it does to scan the ticket.

However, there is no data on whether COTASS has impacted on disorderly behaviour inside the ground and the take-up by clubs of this part of the system has been too low to assess whether COTASS is yet proving effective in access control. Nevertheless, the COTASS team continue to encourage clubs to come on board and are hopeful that their efforts will in due course be rewarded.

Relationship Marketing

National ticketing should be providing sales, service and security benefits, although reduced attendances suggest the sales are not yet being realised. Access control provides mainly safety benefits at a cost to the clubs. Security has been ratcheted up but the benefits profile is looking a bit light for the fans and distinctly lacking from the clubs' point of view.

Yet the Club Card database provides an ideal source of data for clubs to enjoy increased revenue benefits linked to increased customer service. Knowing your customers is an essential part of merchandising and marketing campaigns but very few clubs seem to have yet appreciated the ways they could use the information to increase revenue. Why not send everyone who attended 75% of last season's home games an invitation to buy a season ticket and replica club shirt? Why not telephone everyone who has not renewed their Club Card and find out why? The possibilities are numerous.

Clubs can also give customer benefits by using Club Card as a loyalty card. For example, FC Volendam have negotiated a range of discounts for their cardholders at retailers in the local area. FC Den Bosch have done the same thing, only the fans enjoy twice the benefits. They get a discount at the retailer, and if they spend a certain amount of money, they get a free ticket for the match. Again, however, the clubs have been slow to pick up on the opportunities here.

Overall, then, it is still early days for COTASS. The national ticketing network is a great idea which could surely be adopted by other sports and other countries. The access control system is under-developed and contains several (deliberate) loopholes. Even if the social costs of closing these are acceptable, it is questionable whether COTASS will do anything to prevent violence outside the stadium. The Club Card concept again is an excellent idea, certainly capable of adoption elsewhere, and which would surely enjoy far greater success if more clubs developed the concept as a relationship marketing and service tool.

3.3 – COMMENTARY BY STEVE FROSDICK – THE DUTCH 'COTASS' PROJECT

This integrated system reported on in 1997 was in many ways ahead of its time. The use of database technology to support ticketing and customer relationship management is now commonplace within UK sports grounds. Various 'smartcard' systems have also been introduced to support ticketing, access control, and (in some cases) to provide 'rewards' for fans. For example, season ticket holders at several football clubs use their 'smart' season ticket card to activate the turnstiles, which thus do not require an operator.

The Dutch found that requiring all fans to have a club card resulted in reduced attendances, which were commercially unsustainable for the clubs. Accordingly, they later relaxed the scheme so that the municipality, the police and the club could consult together and agree on the high risk matches which should be 'club card only', whilst then allowing more unrestricted ticket sales for lower risk games. This is the system which persists to the present day.

The Dutch experience has been mirrored in the UK. A few clubs, for example Millwall, have used 'smartcards' to exercise significant control over spectator access. After serious problems at the end of the 2001/02 season, Millwall required all their fans to apply for an individual 'teamcard' for the 2002/2003 season. Ticket sales were restricted to one ticket per cardholder and admittance at the turnstile was contingent on production of both the 'teamcard' and the ticket. As was the case in Holland, attendances dropped, although the team's poor form was also a factor in this. However many people attend matches with friends or family. Usually one person goes and buys the tickets for everyone in the group. Faced with the requirement for everyone to attend the ticket office, some groups of fans stopped coming – the logistics were simply too complicated. Millwall therefore came under commercial pressure to relax the controls and, for some lower risk matches, then allowed cardholders to purchase more than one seat. For low risk matches, unrestricted ticket sales were later reinstated.

The key point for both countries is of course the fact that the existence of the card scheme still allows the club to restrict ticket sales for high risk matches to card holders only, which is a significant benefit for public safety and order.

The Dutch facility of being able to buy match tickets through national lottery outlets has not been taken up elsewhere, probably because of the technological advances which now allow fans the convenience of purchasing tickets over the Internet. These are then posted out, or, as is increasingly becoming the case, printed by the fans from their own computers using the advanced 'print at home' technology now available.

3.4 – UNTHINKABLE

The original citation for this article is: Frosdick, S. (2002) 'Unthinkable', *Panstadia International Quarterly Report*, Volume 8 Number 3, February 2002, pp. 14-15.

The terrorist attack on the World Trade Centre has prompted venues to reflect on the adequacy of their safety and security arrangements. Steve Frosdick reports on a panel discussion at the recent International Sports Facilities Forum

'Panstadia' was a media partner for the International Sports Facilities Forum held in London on 4-5 December 2001. The Forum was organised by Rob Berlinger from the World Research Group and presented by Steve Cameron from Street and Smith's 'Sports Business Journal'. A wide range of themes were covered in the conference programme, including financing, public/private partnerships, real estate development and technological innovation. It had also been intended to feature fire safety risk assessment and strategies for venue safety and security. But in the light of reactions to the terrible disaster in New York, these sessions were reorganised under the title 'First Principles in Venue Safety and Security: What's Changed After September 11?'

The discussion panel was chaired by 'Panstadia' contributor Steve Frosdick from the Institute of Criminal Justice Studies at the University of Portsmouth. Steve was joined by two panellists. John de Quidt is Chief Executive of the Football Licensing Authority and also chairs the Council of Europe Standing Committee concerned with safety and security at sporting events. Dr Adrian Hay is a Director of Tenos Fire Safety Engineering. These three experts constituted a balanced panel of commercial, regulatory and academic perspectives.

John de Quidt opened the discussion by summarising his own first principles, derived from ten years experience of visiting sports facilities across Europe and indeed the world. He argued that whatever the place, whatever the facility and whatever the sport, the following general principles were universally valid.

- First, there was a need for a co-ordinated approach over facility design and facility operation. There was then a need for a clear vision of who is responsible, both overall and for what.

- To avoid confusion, division or overlap, venues should identify responsibilities and put them in writing.

- There was a need for systems and procedures, including a written safety policy, risk assessments, contingency plans and exercises.

- Proper equipment was required, for example a control room, CCTV and communications.

There was a need for personnel, such as a safety officer with competence, status and authority, and properly trained and briefed stewards.

- Next came structures, including seating quality, viewing standards, maintenance and checks.

- Then there were questions of the environment and who you wanted to attract, Considerations would include the quality of welcome, catering and toilets and the needs of disabled persons.

- Finally there were rules and regulations; such as the permitted capacity, the safety certificate, the criminal law, venue regulations and general oversight by the authorities.

John emphasised that these general principles were unaffected by the events of September 11. The venue management should already have been taking whatever measures were necessary to ensure the reasonable safety of spectators. While particular risks might vary over time and between facilities they all required the same coherent approach.

Adrian Hay echoed these first principles and agreed that September 11 was unlikely to have implications for sports facilities. But he did feel the New York disaster highlighted two very important issues. First, designers and regulators do strive to achieve a building where the residual risks are 'as low as is reasonably practical' (the ALARP principle). But no building can be designed to be one hundred per cent safe. Second, good management is just as critical as good design – all the physical design precautions in the world will be useless if there are no emergency procedures in place.

He went on to warn against any 'knee-jerk' reaction to the disaster, emphasising the need for a risk assessment based approach. This underpinned the 1997 fourth edition of the UK Guide to Safety at Sports Grounds – the 'Green Guide' – which was regarded world-wide as a best practice benchmark for the safe design and operation of sports facilities. Would it really now be reasonable or practicable to redesign and rebuild venues to withstand the impact from terrorists flying an airplane on a suicide mission? Of course not – but there were difficulties in trying to establish what did constitute a reasonable level of safety and security. Risk assessment requires consideration of not only the cost of protection but also the public perception of risk. Whilst responses to September 11 clearly indicate increased public aversion to low probability high impact events, nevertheless the health and transport industries have to put financial values on human life to make decisions. Facility designers have to make similar choices, ensuring that good design takes full account of the hazards and strikes a balance between the costs and benefits of risk mitigation.

Steve Frosdick agreed with the need to avoid 'knee-jerk' reactions. He argued that British football had shown the failure of 'legislation by crisis' as public policy. The sorry history of disaster ... legislation ... disaster ... legislation had resulted in a previously piecemeal framework of controls rather than in looking at safety and security in the round. The 1997 'Green Guide' was such good practice because it was the product of leisurely reflection and the consolidation of much other guidance material. It would therefore be a mistake for the sports facilities industry to make fundamental changes to venue safety and security on the back of one wholly exceptional event.

But September 11 had brought two things into sharper focus. First, the owners and operators of sports and leisure facilities remain unequivocally responsible for the safety of their customers. If and when the next disaster strikes, the climate of risk as blame means that people will be called to account for the adequacy of their arrangements for public safety and security. Nothing has changed here, but September 11 has undoubtedly concentrated people's minds on the issues of responsibility, accountability and civil and criminal liability.

Second, there was something cataclysmic about September 11 which has resulted in all kinds of industries wanting to learn lessons from what happened. As one friend from another industry had told him, 'the world has gone contingency planning mad'. The sports facilities industry has a poor record for learning lessons and/or implementing recommendations, both from within the industry itself and from other industries. Notwithstanding the Bradford soccer stadium fire in 1985, cases of spectators being locked in during the event still occurred at venues hosting other sports during the 1990s. Notwithstanding the people crushed to death behind pitch perimeter fences at Hillsborough in 1989, pitch perimeter fences with inadequate exit gates can still be found in some parts of continental Europe in 2001. Outside the industry, project management shows the need to engage with users at the design stage of all kinds of projects. Yet there are plenty of examples of sports facilities where the users have been ignored – venues which look fantastic but which just don't work in practice.

Given this poor track record, why was it that sports venues suddenly wanted to jack up their security arrangements? Safety and security are a risk management problem – maintaining safety and security means identifying all the relevant hazards, estimating the probabilities and impacts, evaluating the risks, and responding to the risks with appropriate control measures. But following events in America, there were examples of venues implementing far more rigorous searching of spectators entering the facility and even of requiring special security markings on catering supplies brought in for banqueting functions. What risks were these control measures trying to address? Would they prevent terrorists crashing a plane on the venue? Or was the intention perhaps to prevent people bringing anthrax spores into the facility?

September 11 has changed both perceptions of risk and responses to risk. Now that the unthinkable has happened, it is no longer unthinkable. People's perceptions of the probability of such events have increased. And the quite extraordinary loss of life means that people's perceptions of the impact have increased too. Concepts of 'massive loss of life' have changed from tens or hundreds of deaths to the horror of thousands killed. Thus responses to risk have been ratcheted up. Proportionate responses based on the 'ALARP' principle have been overtaken by the precautionary principle used when science is uncertain. 'If we can't be sure it's safe, then we don't use it' may work for new and untested drugs, but is hardly sensible for sports facilities.

Steve argued that there were serious problems of potential over-reaction here. Historically, safety had been compromised because of an over-emphasis on security – 'protection from the crowd' rather than 'protection of the crowd'. We need to be very careful that responses to September 11 don't have the same effect. There was a danger that security managers would use recent events as an excuse for pushing through 'pet projects' which could not otherwise be justified. Any such ratcheting up of security might have adverse consequences elsewhere. More fastidious searching could result in entry delays and a risk of crushing on passageways to entry points. Security expenditure unjustified by risk assessment could damage profitability.

By all means let's review our risk assessments and contingency plans. But let's keep our responses to September 11 as proportionate to the risks as we can. Security responses which are disproportionate to the perceived risk undermine safety, ruin people's enjoyment, damage profitability and have negative effects on the neighbourhood. Don't go there.

3.5 – SECURITY PENDULUM

The original citation for this article is: Frosdick, S. (2002) 'Security Pendulum', *Stadium & Arena Management*, June 2002, pp. 9-10.

Steve Frosdick reports on reactions to September 11 in the United States.

Following the attacks on the World Trade Center and Pentagon on September 11 last year, safety and security assumed a new prominence for stadium managers. To the existing risks of disorderly behaviour, crowd crushing and technical or structural failures was added a heightened risk of terrorism. The discovery in the US mail of packages containing anthrax spores provided a further dimension. So the 'Sports Venue Safety and Security' conference presented by the 'SportsBusiness Journal' in Phoenix Arizona in February was a most timely and very well attended event. The 'SportsBusiness Journal' reporter and columnist, Steve Cameron, chaired the conference.

In the immediate aftermath of the attacks, there had been much media criticism of the generally relaxed nature of US security. Take airport security for example – catching an internal flight had been rather like getting on a bus, with minimal checking evident. And that was fine. There was no perception of the possibility of such a large-scale terrorist spectacular with such devastating consequences. Passengers wanted cheap tickets and minimal inconvenience to their journeys. Airlines had trimmed costs and margins in order to compete. It's easy to be wise with hindsight, but low-key security was the natural and understandable result of commercial pressures and consumer demands.

There was general acceptance among the conference delegates that the pendulum had swung too far in the relaxed direction. Summarising the discussions, Steve Cameron noted that, 'America had become complacent. September 11 caused a re-evaluation and increase in security.' Panellist J. Isaac from the Portland Trail Blazers/Oregon Arena Corporation commented that, 'the key early decision was that we were going to materially improve'.

How quickly the pendulum had swung to the other extreme! Public, media and security reactions to the cataclysmic nature of the attacks combined to palpably increase the perceived risks and to demand visible action to provide reassurance that 'something was being done'. As one delegate put it, 'Security used to be about stopping people sneaking in – now it's all about keeping dangerous things out'. Governing bodies such as the National Basketball Association had issued extensive new security recommendations. And speakers and delegates reported a wide range of new and upgraded security measures. These addressed three key areas: the risk of something crashing into the facility; the risk of a bomb being brought in; and the risk of a biological attack.

Milton Ahlerich from the National Football League and Kevin Hallinan from Major League Baseball explained the quite extraordinary precautions that were now being taken for events designated as having national security importance, for example football's Super Bowl and baseball's World Series. Huge numbers of personnel from Federal Agencies such as the Secret Service would be drafted in to supplement local police and security personnel. Very detailed planning would take place. During the Super Bowl, an air reconnaissance aircraft had monitored the skies and fighter jets had patrolled – presumably ready in the last resort to shoot down any plane with aggressive intentions.

Such extreme measures clearly can't apply to every event at every venue. But the Arizona State University stadium is on the flight path for Sky Harbor International Airport; and because of concerns about public paranoia, General Manager Tom Sadler had been able to get non-commercial flights suspended during matches at his facility.

Other delegates reported installing additional bollards, barriers and 'Jersey barricades' around various facilities. These engineering responses were not pretty to look at, but were perceived as necessary to prevent rogue vehicles being driven into the stadium or its fabric.

Many venues have taken the opportunity to network with other venues to exchange ideas and measures to lock the venue down were evident everywhere. There was a shortage of bomb dogs, which were much in demand at airports as well as for sniffing venues, vehicles and rigs for explosives. At some places, all VIP and players' bags, concert equipment, etc. were searched and tagged on arrival at the venue. Careful searches were being carried out before the venue was opened and any non-tagged items were treated as suspicious. One delegate was encouraging more rigorous searching by hiding 'swag' such as $100 bills for the search staff to find. On non-event days, visitor access was very strictly controlled, with some venues allowing visitors by prior appointment only and others requiring visitors to leave their driving licences at reception.

Staffing levels had been dramatically increased and security responsibilities extended. Rick Humphrey from Talledega Speedway reported 'We staffed up the gates and security personnel, and also used the ticket takers as first line security. This allowed us to increase security but still get everyone in before racing started'.

And it was substantially increased access control that provided the main security tool. Sports patrons who had been accustomed to bringing bulging cooler bags into venues found that things had dramatically changed. Many places limited patrons to one soft-sided bag with certain maximum dimensions, and each individual bag was carefully searched. Airport-style metal detectors were being used in some venues and hand-held wands in others. In some cases, guests were banned from bringing in anything whatsoever. No storage or disposal facilities were offered and so, for example, a woman carrying a handbag or large purse could simply be turned away. One might have thought there would be uproar at some of these changes, but delegates reported that the public had been very supportive. The venues had made considerable efforts to publicise the changes, with extensive publicity campaigns – television, radio, newspapers and the Internet – and new signage to make the position clear. The consistent application of policy at all entrance points had also been key. So long as they didn't see others being allowed to take bags in, people would accept the ban.

Blanket application of policy also provided protection against the threat of litigation if things went wrong and something did get in to a facility. One speaker commented that he was troubled when he saw the 70 year old lady who had been a season ticket holder for ten years having her handbag searched, but any exercise of discretion provided a stick for an attorney to beat the venue with.

Various technical advances were debated, including the use of facial recognition software attached to CCTV systems used to scan the crowd for known terrorists. However, as several delegates pointed out, would a stadium manager be willing to evacuate thousands of people from a venue on a possible false alarm from the software – no!

The anthrax incidents had prompted some quite elaborate procedures for handling deliveries. Some venues had built a separate delivery building away from the main venue to receive all parcels, mail and other deliveries. Many venues were scanning all incoming items for biological contamination using staff wearing special protective clothing. And at some facilities, security staff were mounting 24-hour guard on air intake ducts to prevent anyone throwing anthrax spores into the heating and ventilation systems.

Being the sole European at the conference, I hesitantly raised the question of whether some of these changes were a little excessive, if not surreal. Security costs must have been ratcheted up considerably and I wondered how long corporations would be willing to continue to pay. Many of the measures were draconian and I speculated that the public might become fairly irritated fairly soon. Delegates were willing to concede that the pendulum might have to swing back a little, however there was an overwhelming view that September 11 had changed the world forever.

Three months on from the conference, I made contact with a number of the delegates to enquire on the current state of affairs. The general view was that security had increased but that public support had been retained. According to Steve Cameron, 'The security posture in the US is still very brave in public. Nobody wants to say out loud that economics and the staggering long odds of a terrorist attack have allowed them to scale things back a bit. It's fair to say that most operations people really have raised the bar a bit, all around. September 11 forced them to evaluate security and safety issues, which was all to the good, and some positive things have come from that. However, the foxhole mentality is gone. Exceptionally tight security – almost military in nature – will continue to exist for high-risk events like the Super Bowl. On a day-to-day basis, I believe venue managers have satisfied themselves that things are more in order (they use the phrase 'heightened awareness' a lot) and spectators – even in places where they must pass through metal detectors or are forced to arrive without bags – haven't seemed to get upset. Attendance has not dropped in any sport at any level. The bottom line is that there is still lots of brave talk about security, but after the first rush to improve things and look over procedures, a few proper changes were made and that's the way it's stayed. In the end, as you would expect, common sense will prevail.'

Figure 3.5.1 Baseball fans experienced a very relaxed environment prior to 9/11. Bag checking is now standard security practice.

3.4 AND 3.5 – COMMENTARY BY JIM CHALMERS – VENUE SECURITY AFTER 11 SEPTEMBER 2001

Both articles written by Frosdick in 2002 need to be taken in conjunction since they provide a comparative study of both a UK and US perspective of venue security in the aftermath of the Twin Towers atrocity in New York on 11 September 2001. In article 3.4, Frosdick writes about a panel discussion he chaired at a Conference in London in December 2001. In article 3.5, he writes of his experiences when he attended, as a delegate, a sports venue and security conference in Phoenix Arizona in February 2002. Interestingly, he was the only European delegate at the conference. Both articles provide an interesting comparison of views on venue security but without judging who is right or who is wrong in their approach to the issue.

At the London conference Frosdick was joined by John de Quidt, arguably the leading UK expert on the subject of safety at sports grounds, and Dr.A. Hay, an expert in fire safety engineering. The panel examined stadium operations from a commercial, regulatory, design and academic perspective in the aftermath of September 11. De Quidt's first principles in venue safety and security should be framed and on the walls of offices occupied by all venue managers, particularly those involved in the areas of stadium safety and security, as an aide memoire of their onerous responsibilities and accountability.

Overall the panel came to the view that the effects of September 11 should not have any major impact on the general principles of venue safety and security, arguing that management should already be embracing these principles on the basis of hazard identification and risk assessment. However the panel did not pursue what that level of risk was and how it should be managed and this contrasts with the American response to the issue. The panel were in accord that the security threat and response should not stem from a 'knee jerk' reaction, which could turn a stadium from a public assembly facility into an impregnable fortress. In this regard the article presented no new insights into the processes of hazard identification and risk assessment related to the threat of terrorist activity but instead questioned just how far venue management should go in making venues less susceptible to terrorist activity, but with no answers as to how this could be done.

By contrast the US conference examined increased venue security measures on a practical basis, as distinct from a theoretical standpoint, which I would argue formed the substance of the London debate. Frosdick, perhaps unwisely, questioned whether some of the US measures were excessive, given the attack on their homeland. The US reaction was understandably the opposite with public support for the enhanced security measures which had been introduced. They argued that September 11 had changed the world forever and that this included how sports venues were managed and operated. This view is shared by the British

Prime Minister, Tony Blair, who said, 'If the 20th century scripted our conventional way of thinking, the 21st century is unconventional in almost every respect. This is true also of our security. The threat we face is not conventional'.

Given that view, perhaps the London Conference should have gone outside the normal safety and security conventions in assessing how UK stadia should respond to the terrorist threat in the post September 11 era.

The London conference debated issues of increased security bringing a perceived risk of undermining safety, ruining enjoyment, damaging profitability and disturbing the neighbourhood. The US experiences were about providing reassurance to the public. There is a complex balance to be struck here.

The London panel placed great emphasis on the need for venue management to review their security risk assessments and contingency plans but three years after the conference do we know what has actually been done? One significant step was a conference and demonstration arranged by Staffordshire County Council Emergency Planning Unit and Stoke City FC. The event was held in July 2004 to discuss how sports grounds should plan their response to a chemical, biological, radiological and nuclear terrorist attack (a CBRN incident). Since the conference was aimed principally at football ground management the response was very disappointing with only one football Director putting in an appearance. Even the response from safety officers was poor. There were many lessons to be learned, particularly on contingency planning.

As a speaker on contingency planning at football, cricket, rugby and horseracing safety officer courses, I am concerned that most venue contingency plans do not include an initial response to a CBRN incident. In terms of planning how many venue managers have sought advice from their local Police Counter Terrorist Security Advisor (CTSA)? I would suggest not many. In building or developing a stadium what physical anti-terrorist measures have been included in the design stages? Again in my experience not many. How many safety officers include in their event day preparations the risk of a possible terrorist attack ? Again I would suggest not many.

Perhaps because of my interest in the subject even my small club, Kidderminster Harriers FC, has taken the panel advice in planning for the unthinkable. Frosdick's article quite rightly raises concerns about how far one should go in the security of a sports venue but it would appear that the very basics of security reviews and contingency planning for the threat still have a long way to go.

Since neither the fourth edition of the 'Green Guide' or the Football Licensing Authority 'Guidance on Contingency Planning' mention the threat of a terrorist attack, this in some quarters may be all the excuse they need to do nothing. The London panel spoke of their fears of going over the top with the venue security response. I would argue on the contrary that not enough is being done to plan

Safety and security at sports grounds 107

and protect our sports grounds against what the Prime Minister describes as the unconventional threat. In the post Hillsborough era much emphasis has been placed on the need to develop a safety culture at sports grounds. In article 3.5 Frosdick examined the 'Security Pendulum'. If the security threat is to be taken seriously then the pendulum needs to swing a bit more in the security direction.

Figure 3.5.2 High-profile international events, such as Athens Olympic Games, are considered potential terrorist targets. The Helleniko site pictured here contained many sport venues inside one security cordon.

PART IV – SEGREGATION

Introduction

Part IV, comprising five articles, deals with issues relating to segregation from both the physical and management perspectives. The articles examine various methods of segregation and in particular the use of fences as a means of spectator control. A case study examines present day segregation methods at football matches, and the final article examines player behaviour and its effect on the crowd.

4.1 and 4.2 – Segregation From the Field of Play

These two articles were written within a few months of each other in 1997. Segregation methods at various types of sporting events and in various countries are mentioned but the main theme of both articles relates to segregation methods at football grounds and, in particular, the use of pitch perimeter fences. Comment is made about the adverse effects of such fences for crowd safety and comfort, although it is recognised that in some countries this may be the only way of ensuring football matches can be played.

4.3 and 4.4 – Crowd Segregation at High-Risk Matches

Written in 2004, this pair of articles present a two-part case study of crowd segregation methods at fixtures involving Southampton FC and Portsmouth FC. Frosdick provides a unique, independent review of policing, management and physical segregation methods and tactics with an examination of their effectiveness. He makes several recommendations aimed at improving segregation methods. Comment is made that despite a 1991 Home Office Affairs Committee Report recommending that steady progress should be made towards desegregation at football matches nothing has changed in the intervening years.

4.5 – Player Misbehaviour Affecting Crowd Safety

This 1997 article examines player behaviour on the pitch and the effect this can have on crowd conduct and control. Frosdick examines the risks to spectator safety as a consequence of players' actions and questions whether such activity should form part of event contingency planning. Comment is made that some eight years after the article Frosdick's findings are just as relevant today. More drastic measures may yet be needed to curb the extremes of player misbehaviour and misconduct.

4.1 – KEEP OFF THE GRASS

The original citation for this article is: Frosdick, S. (1997) 'Keep Off the Grass', *Stadium and Arena Management*, Volume 1 Issue 4, June 1997, pp. 26-28.

In a modern sports venue it is necessary to keep the spectators away from the area where the event is taking place. Steve Frosdick looks at the problems and possible best practice in carrying out this difficult task.

Spectator Segregation

Spectator segregation is one of the key design and management issues in public assembly facilities (PAFs). In the interests of preserving public order, particularly at football matches, there is a perceived need to segregate opposing groups of spectators, first on their approaches to the stadium, and second within the viewing accommodation. Thirdly, there is the need to segregate spectators from the area where the event is taking place. This type of segregation and the problems posed by spectator incursions onto the event area are common to many sports.

Some examples from recent years illustrate the different types of hazard. The start of the 1993 English Grand National horse race in Liverpool was disrupted by a course invasion by animal rights protesters and the subsequent abandonment of the race was a public relations nightmare. A near disaster also resulted from the mass `friendly' invasion of Silverstone Circuit after Nigel Mansell's victory in the 1992 British Grand Prix. The thousands of motor racing fans who streamed onto the track had forgotten that the other cars were still completing the race! The on-court stabbing of tennis star Monica Seles during a match in Germany was widely reported and resulted in the revision of the security arrangements for many tennis tournaments. And who would be a referee, given the regular reports from around the world of these poor officials being chased off the pitch or even attacked by supporters?

Conversely, the spectators also need to be protected. An extreme example is provided by Eric Cantona of Manchester United who drew banner headlines for his karate style attack on an abusive spectator. More commonly, spectators near the front of the viewing areas face the risk of injury through being struck by the ball, kicked by a horse or even, as in the case of the TT races on the Isle of Man, hit by a motorcycle crashing off the circuit. These cases give rise to concerns about the liability of PAF's operators to pay damages for the injuries sustained.

Finally, within sports such as football, and to a lesser extent in rugby, the need for such segregation is illustrated by the regular occurrence of hostile pitch invasions by large numbers of disorderly supporters. This is not only a serious safety problem, as was seen in the well-publicised incident resulting in the abandonment of an Ireland versus England football match in Dublin, but also a major commercial concern. There is widespread consensus that the resurgence of English football as successful business would be severely undermined by the return of widespread hooligan behaviour, such as opposing groups of supporters fighting on the pitch.

Thus the existence of the hazards, their regular frequency and the adverse consequences of occurrence – for good order, for safe enjoyment as well as for the business – combine together to create an unacceptable level of risk. Control measures are clearly required. Historically, however, the control measures adopted have themselves created a serious risk to public safety and enjoyment.

Barrier Team

In stadia and arenas, the use of physical barriers has been the most commonly adopted control measure. In some sports, for example ice hockey or motor racing, the risk of serious injury from the puck or from debris flying into the crowd is so great that a physical barrier – a transparent plastic screen or debris fence – is clearly essential. In other sports, however, the physical barrier itself has been a contributory factor in disasters. The deaths of 95 English football supporters, many crushed to death against the high perimeter fence, at Hillsborough in Sheffield in April 1989, provides a clear illustration of the potential for tragedy.

In the light of the Hillsborough disaster, there is considerable controversy in Britain about the safety risks posed by high perimeter fences. The official 'Green Guide' to Safety at Sports Grounds, the latest version of which was only published in March 1997, is clear that, for new construction, 'the use of pitch perimeter fences is not recommended for standing areas, and should be avoided in all cases in front of seated areas'. And FIFA, the world governing body of football, in its 'Technical Recommendations and Requirements for Construction or Modernisation at Football Stadia' also states that, 'Ideally, the playing area of a stadium should not be surrounded by security fences or screens'.

Yet the caging of supporters behind high metal fences is a practice still dangerously prevalent at stadia throughout the world. Indeed the French government decided in February 1997 that the fences present in the existing grounds to be used for the 1998 World Cup will not be taken down. The new Stade de France national stadium, being built in Paris, will also have such fencing. These decisions were greeted with grave disappointment in England, particularly as the organisers were known to have been committed to making the necessary improvements in safety management and y to allow for the removal of the fences. In the particular interests of safety, it must be hoped that the French heed the essential recommendations in the Green Guide concerning such fences. These address such matters as gates and openings, the removal of spikes and overhangs, the provision of cutting equipment, lateral gangways and loadings.

Alternative Assurances

So if high fences in front of viewing areas are bad practice, what then are the alternatives? The 'Green Guide' refers to 'high profile stewarding, clear signs, regular public address announcements, and/or the construction of lateral sunken gangways'. FIFA mention police/security personnel, dry moats, removable transparent sections of fence and a seating configuration which puts the front row spectators at a height above the arena which would render intrusion into the playing area improbable if not impossible.

These are all good individual ideas but what I have found in my own research is that the most successful approaches seem to be those which combine an adequate physical barrier with a variety of other techniques to maximise the message. So what I want to do now is to give some examples of good practice, and then conclude with a case which illustrates my argument that the more you tell spectators in different ways that you don't want them to go onto the event area, the less likely it is that they will do so.

Some stadia, including several of the most recently built stands, have only advertising hoardings or a low rail to segregate spectators from the activity area. This is inadequate, and a fence or wall of 1.1 metres, incorporating adequately staffed emergency exit gates, in my view represents best practice. This provides a substantial barrier without compromising either viewing quality or emergency egress. The provision of gate stewards allows for public reassurance, crowd monitoring and immediate reinforcement of the barrrier where necessary. As well as the dry moats found at European stadia such as Utrecht in Holland or Munich in Germany, alternative barriers include the horizontal elastic fence patented by Wembley stadium. The great advantage of this fence is that it does not create restricted views. It consists of several stretched horizontal wires running parallel to the pitch perimeter and creates the effect of trying to walk across a sagging mattress.

Where appropriate, physical barriers can be reinforced with different intensities of police and/or steward perimeter cordons. I went to one Grand Prix motor race where, to help discourage a circuit invasion, loose cordons of stewards slowly and unobtrusively took up a position in front of the debris fence during the final laps of the race. At a derby football match I visited in north-east England, a special plan had been prepared to prevent a pitch invasion at the end of the game. With two minutes to go, the pitch was enclosed by a solid cordon of 367 police and 265 stewards, all standing flush with the pitch perimeter wall. On the final whistle, 18 police dog handlers and their dogs ran out onto the pitch and formed a secondary cordon from penalty area to penalty area. The end result was an awesome albeit resource intensive sight.

At another football ground, I saw stewards seated on low stools all around the pitch perimeter. Each had a radio with a headset. They generally faced into the pitch but on receipt of a radio command from the control room or on the scoring of a goal they instantly swivelled round and faced the crowd, standing up if necessary to create an instant cordon. This ritual was demonstrated by the control room Sergeant for my benefit and was indeed impressive.

But barriers and cordons need to be supplemented by spectator education tactics. These include notes on ticketing and in the event programme, the display of warning messages on electronic scoreboards and video screens, repeated public address announcements, clear warning signs and the rigorous enforcement of ground regulations and legislation. Unauthorised entry onto the pitch is a criminal offence at designated British football grounds.

I went to one London ground where there were signs in large red letters all along the back of the perimeter advertising hoardings warning spectators that if they passed that point they would be ejected from the ground. When the away team, who were losing 5-0, scored a consolation goal, one person jumped up and stepped over the hoardings, arms aloft. As he stepped back, he was arrested by stewards and immediately taken off to the police room where he was later ejected. At many grounds he would have been allowed to go back to his seat. The control room staff explained their philosophy to me. The signs were unequivocal. If anyone disregarded them, they were always arrested and ejected as an example to others. It was felt that any failure to make an example of the first person to encroach could encourage others to be bolder later on in the game. The home supporters had learned that the stewards and police meant business and hardly ever encroached in this way. If the away supporters were unaware of the home club's reputation in this matter they were soon very clear about it.

Let me conclude with a case taken from a football cup semi-final in the north of England. The police briefing addressed the possibility of a celebratory pitch invasion. This was a very sensitive issue. A very disorderly pitch invasion had occurred only seven days earlier at another ground and it was felt that any pitch invasion would be bad press for football. The police and club had discussed the risk and decided that appropriate announcements and steward deployments would be made. It was agreed that the players would do a lap of honour if they won. The atmosphere at this match was one of extraordinary excitement. The away team led 1-0, then later on the home team scored twice in quick succession. After the second goal and a minor attempted pitch incursion, the Safety Officer redeployed stewards from the away end to the front of the home end. Police reserves were also deployed. Towards the end of the game, a considerable number of stewards were slowly deployed on the perimeter track all around the ground.

The public address announcer was contacted and asked to make announcements during gaps in play, which he did on several occasions. Additionally, messages appealing to the crowd were flashed on the electronic scoreboard. The crowd were being asked to stay off the pitch, with the promise of a lap of honour by the players if they complied. The away supporters melted away quietly after applauding their team. But it seemed that all the home supporters stayed behind. The atmosphere of joy was unconfined. The team did a lap of honour to a great reception. As they left the pitch, the manager sprinted over to the home end and waved. The fans went even more ecstatic. But there were no problems. These measures taken were successful and there was no pitch invasion.

Figure 4.1.1 Fences keep the crowd off the pitch. But how safe are they?

4.2 – ON THE FENCE: BALANCING THE RISKS

The original citation for this article is: Frosdick, S. and Smith, B. (1997), 'On the Fence: Balancing the Risks', *Stadium and Arena Management*, Volume 1 Number 5, August 1997, pp. 30-32.

High on the agenda at the European Stadium Managers Association Conference in Barcelona was the use of fences in stadia. Steve Frosdick and Bruce Smith report on this controversial issue.

In deciding the segregation measures needed for a particular facility or event, managers have to balance the risks. The maintenance of public order and the safety of players or athletes and officials seem to demand that something is done to prevent pitch invasions and fighting between rival groups of fans.

Fences seem to offer an answer. At the same time, however, the paramount requirements of spectator safety and comfort create a need for unobstructed views and a means of escape in emergencies. The paradox of fences is that they are both a problem and a solution.

For many people, fences would seem to represent a very bad choice. They are intimidating and have an adverse effect on spectator morale and behaviour. Being caged-in can lead to fans feeling badly treated and this causes resentment, hostility and a souring of their relationship with those in authority. Worse still, the fences can be lethal. Considerably more people would have died in the fire at Valley Parade in Bradford in 1985 if there had not been ready access to the pitch, a fact that is driven home even further when videos of that tragic day are reviewed.

Had there not been fences at Hillsborough in Sheffield, then it is possible and indeed probable that many of those who died on that tragic afternoon in 1989 would still be with us today.

The argument is that, without fences, the risks involved in a pitch invasion are much less than the risks to general safety with fences in place. Fence structures can also impede enjoyment of the event because spectators cannot enjoy watching sports action when their view is restricted by pitch perimeter or radial segregation fences.

Safety First

Everyone is entitled to the same degree of protection but it is much easier to get the officials and players quickly off the pitch than it is to evacuate spectators trapped behind a fence.

Fences threaten safety and spoil the view. The logic for their removal seems overwhelming. FIFA have recognised this with their insistence that fences should now begin to be taken down.

Indeed fences seem such a bad idea, that one wonders why they were put up in the first place. There have been only a few recorded cases of players or officials being killed by hostile pitch invasions. And these have been in South Africa, with no known cases in Europe, and certainly none in England, where the fences appear to have originated.

Fences seem almost unique to football grounds and the worst examples can be seen in Argentina, where most of the major soccer grounds are surrounded by fences that may reach 20 feet or more into the air and are often populated by spectators who have climbed up them.

In rugby union, it appears to be a tradition for fans to invade the pitch en masse at the end of the match and nobody thinks anything of it. It is the same with cricket and indeed the major outdoor American sports. Occasional streakers have brought amusement to a variety of stadium or arena events. Sometimes the excitement proves too much and exuberant spectators spill onto the pitch in celebration of points scored or a race won. But these are minor nuisances rather than serious threats.

In fact it appears that fences in English football were erected in the 1970's largely at the instigation of the police, not to stop attacks on players or minor pitch incursions, but as a way of preventing rival groups of hooligans from running the length of pitch to attack each other's 'ends'.

Hooliganism was largely the English disease, yet within a few years, fences were introduced in other countries. The question is why? It's certainly the case that stadium owners have never been known to spend money without a good reason. There are also certain countries where it would be an act of extreme folly for the players to take the field without a substantial barrier between them and the fans. Turkey in particular provides a very good example of where fences are required to keep rival supporters apart and keep the fans off the pitch.

Whilst it is clear that there may be certain matches in certain countries for which fencing, on balance, represents a solution rather than a problem, nevertheless, none of the ESMA speakers or delegates had a satisfactory explanation for the widespread general introduction of fences at football grounds.

Comfort Factor

This is a key point for anybody who insists on retaining fences as a general precautionary measure. I recently visited one English football ground, much of which had been magnificently rebuilt, and was surprised to find that the away supporters' enclosure was still fronted by a hideous fence. The safety officer told me he had kept the fence because it made him feel more comfortable. And the French Ministry of the Interior seem to be taking the same line with their insistence on keeping fences for the 1998 World Cup. But why? Where, one might ask, is the evidence of a threat so strong that it tips the scales in favour of compromising people's safety and enjoyment?

The French decision, flying in the face of FIFA's wishes and against the desires of the organisers, seems to have been the result of a political battle won by a macho police force. Worryingly, the retention of fences is only part of the story. The French police are reported to be refusing to wear high visibility jackets because they look effeminate, declining to display identification numbers for unknown reasons which can only be considered sinister and insisting on mounting an obtrusive control-oriented security operation. (Having been on the receiving end of one of their totally unprovoked CS-Gas attacks in the Parc des Princes, Bruce Smith can vouch that the above measures mean that individual policemen cannot be identified).

The general consensus is that the fences will be in position for the forthcoming World Cup Finals, but there is still time for that to change, with some encouraging signals being given. The recent change of government in France has created a political opportunity to review the decision and alternative barriers are still a possibility. The World Cup organisers visited Wembley stadium for an England match in April and took a good look at their unique lateral fence. And according to Jean-Christophe Giletta from the Stade de Frence, the new Paris venue intends to keep its options open by installing the same type of folding fence as used at Manchester United, a fence which measures 1.1 metres high when folded down or 2.2 metres when folded up.

Invisible Control

But simply removing the fences without doing anything else tips the scales too far the other way. FC Barcelona for instance have some of the most passionate and expressive fans in Spain. The club wanted to make security and safety compatible with an exciting atmosphere. They made the areas behind the goals all-seater, which they felt reduced the risk of bad behaviour.

They relocated the terraces to the highest points of the stadium and introduced 'human barriers' of security personnel around the pitch perimeter. Joan Gaspart of FC Barcelona felt that the absence of pitch invasions over the last three seasons meant they had got it right, but with hindsight he admitted they had taken the decision to remove their fences without doing enough research. Having visited the Camp Nou, noted its inadequate access control and been surprised at the very poor structural condition and steep rake of its gangways, I was tempted to conclude that they had made their choice and then been lucky.

The point is that, to get rid of the fences, you have to get everything else right. To make security and safety compatible with passion and excitement you have to create an invisible control system. The spectators may feel free, but they are in fact controlled by a range of unobtrusive measures, including CCTV, psychological cues and laws to criminalise unacceptable behaviours.

Big drops can also be a deterrent, along with moats, sunken lateral gangways and 1.1m walls/fences with manned exit gates. The cramped nature of many stadia, particularly those in Europe, makes these difficult to physically implement, although a wide moat around the pitch is a feature of the Amsterdam Arena.

The British government 'Green Guide' to Safety at Sports Grounds, contains much useful guidance on best practice and I covered this in more detail in the June edition (see article 4.1). The success of the Euro '96 competition shows that a better environment and better management can work, but it took the English seven years to get there, and many other countries have not yet started their journey.

Balancing the Scales

Looking to the future, Ernie Walker, Chairman of the UEFA Stadia Committee, has reported that UEFA were reflecting on what needed to be done to make fences unnecessary and that he had written to FIFA to seek to agree some common objectives. The phased approach to the installation of all-seating had worked well, and it might be that a similar approach could be adopted. He would be very surprised if the tender documents for the 2006 World Cup did not include a requirement for there to be no fences, and he hoped that future finals of UEFA and FIFA competitions would only be held in countries where the atmosphere was civilised, the stadia all-seated and the fences removed.

Fences may once have seemed like a good idea, but the scales world-wide are now settling on the side of safety and comfort. How dreadful that there is a lack of real clarity why they tipped the other way in the first place.

4.1 AND 4.2 – COMMENTARY BY JIM CHALMERS – SEGREGATION FROM THE FIELD OF PLAY

Article 4.1 written by Frosdick in 1997 and Article 4.2 written later the same year by Frosdick and Smith examine issues relating to sports ground segregation from both the structural and management of crowds perspectives. Reference is made to the general principles of segregating spectators from each other on the approaches to and from the sports ground and whilst they are in the viewing areas. The issue of segregating spectators from the sport participants is also examined in detail, particularly the debate about the use of pitch perimeter fences as a means of spectator control.

It is recognised in the articles that football on a global scale has created the most debate on segregation methods due to the very partisan nature of the average football fan. Despite much research we are no further forwards in understanding why equally fervent partisan support for cricket or rugby has not spilled over into crowd misbehaviour. This is why segregation is not such a major issue in these other sports with thousands of opposing supporters mingling freely without a desire to end up in physical confrontation.

The articles concentrate on the segregation of spectators from the area of the event and football is not unique in this respect, as Frosdick and Smith explain. Sports such as motorcycling and motor racing can create different risks if the machines are not confined to the race circuit.

However I agree with Frosdick and Smith that it is the experience of football which has attracted the greatest publicity through the use of pitch perimeter fences as a physical barrier to prevent spectators entering the field of play. As to how and when such fences came to be introduced into football grounds in the UK, Frosdick and Smith are not entirely accurate with their description in article 4.2. The facts are that the first high pitch perimeter fence to be installed at a football ground in the UK was erected in 1967 at The Den, the home of Millwall FC. This followed a revolt by referees when a referee was attacked and injured by spectators as he left the pitch at the end of a game between Millwall and Aston Villa. The Football Association fined the home Club £1,000 but referees considered the punishment too lenient and threatened to boycott The Den until they were assured of protection, preferably in the form of wire fencing separating the crowd from the pitch. The Club reacted by initially fencing one end of the ground. From this stemmed the proliferation of high pitch perimeter fences at football grounds throughout the UK as the norm for controlling spectators at football matches. Interestingly, when Millwall FC moved to the New Den, it was amongst the first new stadia to be constructed without high pitch perimeter fences.

The installation of high fencing did not however stop at the pitch perimeter, with high fencing and overhangs introduced into the viewing areas to segregate home and away fans from each other. In many grounds therefore, the fans (mainly the visitors) were corralled on three sides with high fencing. To some the erection of fences was felt inadequate. In 1985, Ken Bates (then Chairman of Chelsea FC) reportedly advocated that they should be electrified to stop fans climbing on them to gain access to the pitch. Fortunately his argument did not find favour but it demonstrated just what one Club Chairman felt about his paying customers.

As someone who policed football grounds in the Midlands and saw the introduction of fences, the first impression was that their effects worked in two ways. Firstly, they did help keep fans from going onto the pitch but they did not stop it all together. Secondly was that communication between the police and fans became almost non-existent as the physical barrier became a barrier between them and us. This helped breed a culture of alienation of the fans and created the belief that if fans were caged as animals then they were likely to behave like them.

The fans also felt that the caged areas were their domains to do with as they pleased and any attempt to enter that territory by the police was met with both aggressive physical and verbal abuse. At Birmingham City, for example, a caged gangway had to be provided as a means of accessing the large Tilton Road standing terrace to deal with problems between the home and away fans in this area. I have no doubt that the provision of segregation fencing was a major contributory factor in the escalation of poor crowd behaviour in football grounds in the 1970s and 80s, a lesson that other countries have still to learn from.

In the aftermath of the Hillsborough stadium disaster, as the police commander at Aston Villa FC, I ordered that the pitch perimeter and segregation fencing to be reduced to 1.1m, a request the Club readily agreed to. Many of my fellow football police commanders questioned my judgment, as did many of my officers who feared for their safety as the fences came down. However the reverse proved the case. The home fans, in particular, started to speak to my officers and me again after years of hurling abuse at us through the fencing. Whilst I did not consciously think of it at the time, it proved the point then, and still does today, that fences can control crowds but it takes people to manage people. I lived through the Hillsborough stadium disaster and agree with Frosdick when he says that the high pitch perimeter fence in front of Leppings Lane was a contributory factor to so many deaths with fans being unable to escape forward, and their rescue impossible due to the inadequacy of the so called emergency gates.

Frosdick and Smith question why other countries, particularly those in Europe, have yet to follow the UK example by bringing pitch barrier controls down. Many countries would like to but had it not been for the Hillsborough stadium disaster the question to be posed is whether we would still have high pitch perimeter fences in the UK today. As it is, only one senior football club (Shrewsbury Town FC) still retains one high pitch fence at its stadium. I am at a loss to understand why the club, the Football League, the local authority and the Football Licensing Authority still feel in 2005 that this type of control remains necessary.

I do not think therefore that the UK should sit in judgment of other countries but instead show by example how much better the enjoyment of the game of football can be in the absence of high fences. As an FLA Inspector, I remember the experiences of Euro 96 at Villa Park when visiting fans from Europe could not praise too highly the experience of sitting so close to the action and not having to view the game through a wire mesh.

Frosdick and Smith do accept that in some countries football matches could never be played without the security provided by high fences or moats around the perimeter. They mention Turkey and the fences 20 feet high in Argentina. The South American experience came home to me in 1996 when Jim Froggatt (a fellow FLA Inspector) and I were sent by the Foreign Office, at the request of the government of Guatemala, to assist in the investigation into the cause of how 88 football fans were crushed to death in the Mateo Flores stadium in Guatemala City.

Fencing, topped with barbed wire, surrounded the entire stadium. When we questioned the football authorities and police as to whether this control was really necessary they made it clear, in no uncertain terms, that if that style of fencing was not there, then neither players nor officials would set foot on the pitch for fear of their life and limb. The police also advised that in the event of any crowd trouble on the terraces they would not go into the crowd for fear of attack. Any crowd intervention was left to the military.

In such circumstances it was clear to us that high fencing was necessary, which explains why we did not recommend removal of the fencing. Incidentally, the pitch perimeter fencing was not a significant contributory factor to the disaster as was widely reported in the media at the time when comparisons with Hillsborough were being made.

Frosdick and Smith quite rightly speak of a balance having to be struck between safety, security and comfort and I agree with them on how this balance has been reached in the UK. It is not enough just to take the fences down since this in only part of the raft of measures introduced in the UK to make our stadia a better environment in which to watch football in the post Hillsborough era. The most significant of those was the introduction of all seated stadia in the top two divisions. However improving the physical environment in not enough in itself and the change in emphasis from policing to stewarding has also been a significant factor. Sadly, at certain high profile fixtures, the presence on the pitch perimeter of large numbers of police officers in support of the stewards to prevent pitch invasions remains an operational necessity, but this must always be preferable to a return to high fences. Thankfully such scenes are rare with pitch perimeter security at most games being left to well trained and competent stewards.

Frosdick and Smith do not mention the legislation introduced in 1991 making it a criminal offence at a designated football match for anyone to enter the field of play without authority. This legislation came about due to the inadequacy of the existing law to deal effectively with pitch invasions at the time when fences were coming down. Not only is it punitive, but as Frosdick and Smith say it is also a very effective deterrent. How effective the legislation is can be judged from the football arrest statistics for the 2003/04 season when out of a total Premier and Football League attendance of 29,197,510 there were only 248 arrests for pitch incursions compared to 336 the previous season. Many of those arrested ended up with banning orders for up to three years. Perhaps had this legislation been considered in 1967 following the Millwall incident then high fencing as a means of controlling spectators might not have been necessary.

The articles written by Frosdick and Smith eight years ago still raise valid arguments about segregation which are sustainable today. Segregation in football grounds is sadly still essential to ensure the comfort and safety of both sets of fans. The passage of time has however shown that high fences are no longer a requirement in the modern stadium, given the changing stadium environment, improvements in CCTV surveillance and effective stewarding. The absence of fences in the UK is the envy of most countries and is something we can feel rightly proud of. Hillsborough taught us a lesson about high pitch perimeter fences and they should remain where they rightly belong, which is as part of the history of UK football. When and if certain other countries can emulate our example remains to be seen.

4.3 – A TALE OF TWO CITIES: PART I

The original citation for this article is: Frosdick, S. (2004) 'Tale of Two Cities – Part I', *Stadium & Arena Management*, August 2004, pp. 23-24.

Steve Frosdick presents a two-part case study of crowd segregation for the high-risk derby matches between Southampton and Portsmouth in the English Premier League.

On the south coast of England, the 2003/2004 football season provided three fascinating local derby matches between Southampton and Portsmouth. The two sets of supporters do not like each other – something to do with both cities having large dockyards. So as with many other local derbies, these fixtures had a high risk of crowd violence.

This article presents a two-part case study of the segregation arrangements for the matches. Part one begins by reminding us what segregation is and then introduces the two stadia. It goes on to set out various incidents at the first match in Southampton. Part two (article 4.4) will continue the story and then draw out some general conclusions.

There are two types of segregation. First, we have to prevent spectators getting onto the field of play. This is to prevent attacks on players and officials and to minimise disruption from protests or celebrations. Second, we have to keep the two sets of supporters apart, both inside and sometimes outside the stadium. This is to prevent injuries, damage and public disorder.

Segregation is achieved in two main ways. First are physical means such as fences, hoardings, moats, signage and 'sterile areas'. Second are management methods such as netting, publicity and cordons of police and stewards.

This case study focuses on the second type of segregation – keeping the fans apart – for the matches at the Friends Provident St Mary's Stadium in Southampton and at Fratton Park in Portsmouth. These are two very different grounds, built over 100 years apart.

The St Mary's Stadium was built in 2001 and has a capacity of 32,000. It is set in an industrial area near the city centre and is surrounded by a wide external concourse. Internally, it is a modern and attractive bowl of one single tier with all the seats under cover.

Fratton Park, by contrast, is an old-fashioned football ground, built in 1899 beside some railway sidings. Its main entrance is in a residential street. It has a rough car park on one side and is land-locked by housing and industry on the other three sides. It has no concourse, but an external alley on two sides. It holds 20,000 fans in four separate stands.

Figure 4.3.1 Spectators mix on the exterior concourse at St Mary's Stadium, Southampton.

At Southampton, the away fans are accommodated in a self-contained zone. They have their own turnstiles giving access to an internal concourse with excellent facilities. The seated area is segregated from the home areas by a low fence on one side and by a dense black net on the other side. The net can be moved depending how many tickets have been sold to away fans. Unusually, the liveliest home fans are located in the blocks adjacent to the away fans rather than at the opposite end of the stadium.

At Portsmouth, the away fans have part of one end of the ground, at the opposite end to the noisiest home fans. There is seating but no roof cover. The turnstiles are adjacent to turnstiles used by the home fans. The facilities are poor. The away seating is segregated by a wooden panel on one side and by a small net on the other. As with Southampton, the location of the net depends on the number of away fans.

The first match at Southampton was a Cup match held on a Tuesday evening in December. It was the biggest operation ever mounted for a Southampton game with over 400 safety stewards, 120 security stewards and 350 police officers on duty. The match was sold out with 4,250 away fans expected. Because it was the Cup, it was not possible to change the day or kick-off time. This meant that fans had the opportunity to drink alcohol in the city all afternoon and early evening. There is a general view that alcohol consumption can make fans more quarrelsome and indeed the police were called to deal with quite a number of incidents of disorder in the city during the hours before the match. There were reports of Portsmouth fans trying to storm city centre bars and of rival groups of supporters fighting and throwing missiles both at each other and at the police.

The Portsmouth fans arrived for the match in one of four ways. Some fans came by private car and walked to the ground. Sensibly, these fans concealed their Portsmouth colours until they arrived at the away turnstiles. Other Portsmouth fans arrived by train but these seem to have come early and gone into the city to find a drink before walking to the ground in groups, monitored by the police.

Southampton provide a 'park and ride' facility for away fans about five miles west of the city. Here the away fans can leave their cars and travel to and from the ground by shuttle bus. Other fans travelled on a fleet of hire coaches which were escorted to and from the ground by the police. These 'away' buses and coaches park in a closed street immediately beside the external concourse and very near the away turnstiles. This has good and bad points. On the one hand, the walking distance from coach to turnstile is very short and it is straightforward for the police and stewards to supervise the area. On the other hand, the external concourse is desegregated and so there are home and away fans milling about together.

The arrival of groups from the buses and coaches was uneventful. In general, they got off, walked up to the turnstiles in groups, sometimes singing their hatred of Southampton, and went in. Some fans were pulled aside by the police, searched and warned about their behaviour. The police officers were very assertive if not aggressive. But the desegregated external concourse was problematic. There were many examples of rival fans abusing each other as they walked past and I saw several incidents of one-on-one violence. For example, I saw a Portsmouth fan – who was fairly small – and a Southampton fan – who was fairly large – walking in opposite directions towards each other. Not a word was spoken, but as they passed each other, the Southampton fan suddenly elbowed the Portsmouth fan in the face. There were lots of police close by, but none of them saw this incident, which was over in a few seconds.

Figure 4.3.2 Police and stewards form two cordons.

Inside the stadium, there were changes to the normal segregation arrangements. Because it was a Cup match, Southampton had to give the Portsmouth fans an extra block of seats. The dense net on the one side of the away zone was extended over eight lines of seats. The low segregation fence on the other side was supplemented by a line of police, security stewards and safety stewards and by a net stretched over five lines of seats.

I was sat in the last line of home seats next to the fence and saw that these arrangements worked well. It is true that there was considerable verbal abuse being thrown backwards and forwards between the two sets of fans, and that a minute's silence held out of respect for Southampton's Club President, who had died earlier in the week, was disrupted by jeering Portsmouth fans. It is also true that both sets of fans stood all the way through the game. But I only saw two violent incidents. First was when Southampton scored their first goal in a 2-0 win. The home crowd surged at the low fence, screaming abuse at the Portsmouth fans, who they could not get to because of the width of the net. The stewards responded well, spreading their arms wide to make themselves big and try and preserve the cordon. The police were less effective. Some just stood there and did nothing but one police officer pushed a Southampton fan backwards, sending him and a number of other fans sprawling. This was an inappropriate response by a frightened young police officer. The police and stewards were crushed against the fence in places. One steward was trampled on, and a few of the police received minor injuries.

The second incident was at the end of the match when there were two episodes of coin-throwing from the Portsmouth fans – in fact one of the coins hit me on the head!

At Southampton, both sets of fans are normally allowed to leave together and the external concourse is not segregated. But after this match, the Portsmouth fans were not allowed to leave for seven minutes and the police cordoned off the external concourses on both sides of the away exits so as to allow the away fans a segregated route back to their coaches. Once the cordon went in, home fans had to walk all the way around the external concourse to get away. This was the first time this scheme had been used. It worked ok, although there was some disorder outside the ground. Some of the Portsmouth fans managed to break out of an exit and some groups of Southampton fans managed to get around the police cordons.

4.4 – A TALE OF TWO CITIES: PART II

The original citation for this article is: Frosdick, S. (2004) 'Tale of Two Cities – Part II', *Stadium & Arena Management*, October 2004, pp. 30-31.

Steve Frosdick continues his two-part case study of crowd segregation for the high-risk derby matches between Southampton and Portsmouth in the English Premier League.

Part one of this article (4.3) outlined what segregation is and introduced the two stadia. It then set out various incidents at the first match in Southampton. This second part continues the story and then draws some general conclusions.

The second Southampton match was later on in December and was a League match which, on police advice, was changed from Saturday afternoon to Sunday at 12 noon. This meant there was no chance of the fans visiting any bars before the game. There was no trouble at this match. This was partly because of the lack of alcohol; partly because the police had even more resources (including police dogs and horses); partly because the Portsmouth fans were successfully held in the ground for ten minutes after the match; and partly, perhaps, because people had already got the tensions of the first meeting out of their systems.

The return League match at Portsmouth in March was also changed to a Sunday but this time with a late afternoon kick-off for television. There were 180 safety stewards and 16 security stewards – considerably fewer resources than at Southampton – and 400 police officers on duty. The match was sold out. Southampton were bringing only 2,100 fans. There were two main reasons for this. First was that they were only offered 2,100 tickets. Portsmouth were not convinced that Southampton would be able to sell the full 3,100 capacity of the uncovered end, reckoning that the city's rough reputation, the ground location and the external geography would put the fans off. Second, the match was on live television, so they did not need to be there to see the match live. It was a major logistical exercise to get the away fans safely to the ground and none of them came early to cause trouble, as had been the case back in Southampton.

We saw earlier that the main problem at Portsmouth is that the away entrance is located next to one of the home entrances at the end of a residential street. There are two significant problems with this location. First, you have to get the away fans to it. Second, it is a very cramped space. Special arrangements had been planned for this match. In particular, it was planned to use one of the alleys behind the ground as a queuing point. So the home and away fans would approach from different directions.

Just over half the Southampton fans arrived under police escort in a convoy of 18 coaches. These set down the fans at the usual location, which is about five minutes walk away. For reasons unknown, at the last minute, the police decided not to bring the fans by the usual and most direct route. Instead, they closed the two alleys and the back of the ground and escorted them in that way. This infuriated the Portsmouth fans, who saw the police allowing the Southampton fans to 'take' the outside of their ground. The fans were then lined up down the alleyway and turned around the corner to access the turnstiles. This worked well, and the fans were admitted to the stadium with little more than some verbal abuse from the Portsmouth fans inside the stadium above the alley. Plus there was a hail storm and they all got soaked.

Figure 4.4.1 Police horses in a street near Fratton Park.

The arrival of the remaining 1,000 Southampton fans was far more problematic. This group had travelled by train and the first problem was that their train was late. The second problem was that there was some disorder on the train and so the police decided to hold the fans at the station when they arrived. The third problem was that some groups of local hooligans – who were not attending the match – came and gathered outside the away turnstiles to lay in wait for the fans who had come by train. These had to be dispersed by a police charge before the fans could be brought down to the ground.

By now, of course, the match had started, and it was not until more than ten minutes after kick-off that the fans arrived at the ground. The fans were not happy. The police had told them that the kick-off had been delayed. This was untrue. And although the police were wearing riot clothing for their own protection, it gave them an aggressive rather than reassuring appearance.

Now the problem was that there were 1,000 fans to get into the ground through a bank of only three turnstiles. At normal flow rates of 660 persons per turnstile per hour, it would take 20 minutes to get them all in. Understandably, the queue became restless and began to press forward, shouting 'let us in'. There was some crushing and a few people were getting distressed. So somebody decided to open the exit gates to use as an additional entry point. The intention was to relieve the pressure in the queue but now the crowd surged forward and tried to rush the gates, which here hastily closed. Eventually, the last fans got in at about 4.30 pm – very cross indeed.

Figure 4.4.2 Segregation at Fratton Park.

Inside the ground, the segregation between the two sets of fans was minimalist. There was a net laid over three rows of seats and a line of security stewards. The stadium manager's thinking was that, whilst there may be abuse and bravado, the fans do not actually want to fight each other. So putting them so close that they could do so means they behave themselves. And it worked. There were no incidents of violence around the segregation netting.

There was considerable and vitriolic verbal abuse between the two sets of fans but only two real incidents of note. One was when a water bottle was thrown at a Southampton player by a Portsmouth fan. The other was when several bottle of urine were thrown at the Portsmouth goal-keeper by the Southampton fans. The police responded with a show of force.

After the end of the match, the Southampton fans were held in the ground for forty minutes. This was for their own safety. There were groups of youths waiting outside the ground calling out 'let them out, we want them'. It took considerable police resources to clear the area and there was considerable disorder between the Portsmouth hooligans and the police. The Southampton fans were then allowed out of the ground but had to wait a further hour before the police had control of the surrounding area and could escort them to the station. There was very considerable disorder in Portsmouth that night – one of the worst incidents of football hooliganism in the United Kingdom all season.

Having now set out the case studies, I want to suggest six lessons which we might learn from them and so draw some general conclusions.

First, it seems clear that segregation is still necessary. In 1991 a UK government report suggested that 'gradual but steady progress towards desegregation should be the aim of police and clubs'. Within the stadium, some limited progress has been made towards this aim in the case of clubs who have desegregated 'family' areas for parents and children. But the potential for serious injury, damage and disorder at many high profile matches suggests that this aim is neither realistic nor desirable. Indeed, for the 2004 FA Cup Final in Cardiff, the police for the first time insisted that, in order to prevent violence, all neutral fans attending the game had to say which one of the finalists they were going to support. This was to eliminate what was described as 'an unpredictable mass in the centre of the stadium, a cocktail of different allegiances'.

Second, I suggest that management methods are preferable to physical means of segregation. Neither Portsmouth nor Southampton have high metal fences and cages in their stadia. Such engineering structures create a harsher and less customer-friendly environment. The use of netting and cordons was more flexible, less intimidating and just as effective.

Third, I'm convinced that the nearer the coaches can get to the away entrance, the easier it is to get the fans in and out safely. It is particularly helpful to be able to cordon off the area between the stadium exits and the coaches.

Fourth, I think the two cases show that mass arrivals of large groups of away fans are more problematic than the staggered arrival of small groups of fans.

Fifthly, it is clear that what happens away from and outside the ground does affect what happens inside. The problems the police had getting the Southampton fans from the station had a direct impact on the management of their entry to the ground. I think that it would be a mistake for stadium managers to take the view that they should worry about inside the ground and leave the rest to the police.

Finally, the two cases suggest that experienced stewards are more effective at 'policing' crowds than inexperienced or aggressive police officers. Higher profile stewarding supported by lower profile policing remains the right way forward.

4.3 AND 4.4 – COMMENTARY BY JIM CHALMERS – CROWD SEGREGATION AT HIGH-RISK MATCHES

Both Articles written in 2004 reflect Frosdick's experiences when he attended, as an observer, the Cup and League fixtures between Southampton and Portsmouth football clubs in December 2003 and March 2004. Due to the history of rivalry between both sides the fixtures fell into the category of high risk – local derby fixtures when both passions and the potential for crowd trouble were at their highest. Frosdick used both occasions to examine in detail both the physical crowd segregation arrangements and those exercised by the police and ground management inside and outside the stadium. His conclusions are presented in a unique two-part case study, which reflects the many strands of activities necessary on a match day to keep rival factions apart without spoiling the enjoyment of the innocent majority.

In this regard Frosdick's articles present a fairly unique set of observations since he was probably the only independent observer at both games with no partisan affiliation to either side or any axe to grind with either of the clubs, the police or the sports regulatory bodies. His background as an ex police inspector used to policing matches in the London area, an author on stewarding and safety management at sports grounds, an academic researcher and lecturer on sports grounds safety and security, and as a supporter of Brentford FC, makes him exceptionally well qualified to independently assess and comment upon what he saw and heard at both games. In this regard both articles are important from the sense that there is no bias in what he says and his comments only reflect his actual experiences. The articles are also important in that they continue his arguments set out in articles 4.1 and 4.2. He examines the theory of segregation compared to the harsh reality of practice and draws some significant conclusions of value to safety officers, the police and also to football supporters.

I would suggest that Frosdick's experiences could be reflected at most high-risk local derby fixtures and certainly reflect what I saw on numerous occasions during my twelve years as an FLA Inspector. When I started with the FLA in 1991 this coincided with a report published that year by the House of Commons Home Affairs Committee on 'Policing Football Hooliganism'. Frosdick in article 4.4 refers briefly to this but I feel their comments on segregation are worthy of more detail. In the report the Committee recognised that the majority of spectators may wish to watch a match with other supporters of the same side, but this should be a matter of choice, not necessity. Some fourteen years later nothing has changed with many grounds having large signs saying that any visiting fans found in the home areas will be ejected. Not much choice there for an independent supporter who fancies watching a game in the home end.

The Committee recommended, as Frosdick says, that gradual but steady progress towards desegregation should be the aim of the police and clubs. This report was of course written when police had primacy in the control and management of football crowds inside the stadium. Safety officers who have the effective control of ground safety management on match days have superseded this. Again fourteen years later nothing really has changed other than that the high segregation fences have gone. Issues discussed in 1991 on escorting supporters to and from the ground, the strict segregation which creates an arena for tribal posturing, chanting and threatening behaviour are all evident in Frosdick's case study. Whilst the 1991 Report argued that desegregation of football grounds must be the long-term aim of all concerned, based on Frosdick's case study we are no closer to this in 2005 than we were in 1991. A sad indictment on the game, the sport regulatory bodies and on football supporters.

Frosdick closes his case study with a quote from the Scottish Police Federation in 1991 that 'high profile stewarding supported by low profile policing' is the way forward. Throughout England and Wales the transition from policing to stewarding inside stadia is obvious. What must be of concern however is the level of valuable policing resources necessary for the fixtures presented in the case study. With 350 and 400 police officers being required, together with all the ancillary staff and logistical support necessary for such major operations, it brings into focus the 1991 comment that segregation is extremely costly in police and finance. So many years later nothing really has changed for the police in Southampton and Portsmouth.

As the police move towards full cost recovery the issue of desegregation as a means of improving crowd behaviour may well have to be brought into focus yet again. In an ideal world I would welcome similar case studies by Frosdick on segregation at high profile cricket and rugby matches if only to illustrate the differences in attitudes towards segregation at those sports by sports regulators, police and spectators.

In Article 4.4 Frosdick raises an interesting issue when he comments that the supporters from Southampton FC were held back in the stadium for forty minutes in the interests of safety. This brings into question whether this could constitute unlawful detention and a breach of Article 5 of the Human Rights Act of 1998.

Overall therefore, despite the millions of pounds spent in stadium redevelopments and in making the venue environment so much more pleasant, and the major investments made in managing the venue facilities so much better, including professional stewarding, we appear no further forward in our attitude towards segregation than we were back in the 1960's when high fencing went up at football grounds.

The fencing may have gone but not the fan behaviour, which still makes segregation necessary. If ever there was a prime example of this then you need look no further than the events after the FA Cup fixture between Everton FC and Manchester United FC on 19 February 2005 when scenes of extreme violence, involving both sets of supporters, resulted in 33 arrests and five police officers injured. As long as incidents like this continue then the scenes described by Frosdick in his case study will also continue. Not a very optimistic note upon which to end, but the likelihood of desegregation at football grounds remains as remote as ever.

4.5 – PROBLEMS ON THE PITCH

The original citation for this article is: Frosdick, S. (1997), 'Problems on the Pitch?', *Football Management*, Volume 5 Issue 3, Summer 1997, pp. 12-14.

There are growing concerns about the adverse effects of player behaviour on crowd control. Steve Frosdick discusses the problems, the associated risks and what is being done to deal with them.

Articles about safety and security in football grounds usually focus on crowd behaviour, management systems or engineering structures. But there is renewed debate within football about the risks associated with indiscipline on the field of play. These impact on safety and order, on atmosphere and on the image of the game, as well as on individual player's civil and criminal liability.

In examining this issue, perhaps we first need to look at the connection between player behaviour and what happens elsewhere. In his autobiographical account of the life of a professional footballer, the former Millwall player Eamon Dunphy, now a respected journalist, wrote 'I've never believed that terrace violence has anything to do with what happens on the field'. But this is a minority view. In their 1990 book, 'Football on Trial', Murphy, Dunning and Williams from Leicester University's Centre of Football Research point out that historically, the largest category of spectator disorder 'resulted from anger at the decisions of the referee or the attitudes of opposing players'. Writing in the Daily Mail in 1993, Brian Scovell drew attention to incidents at Burnley and Southend, when inflammatory gestures by players resulted in crowd crushing injuries and the outbreak of disorder.

So there is good evidence that spectator safety is affected by the behaviour of players and that this is not a new problem. But it certainly resurfaced as an issue last season, with PFA chief executive Gordon Taylor writing to all his delegates and the subject forming one of the key themes discussed at the Association of Chief Police Officers (ACPO) end of season football conferences held in Manchester in June. These conferences were attended by all police match commanders, with representatives from the football authorities, the Football Safety Officers' Association (FSOA) and the Association of Premier and Football League Referees and Linesmen (APFLRL).

What Are The Problems?

The main safety concern seems to be around the way that goals are celebrated. The police accept that there will be exuberant behaviour by players when goals are scored and, of course, they are not seeking to curtain this spontaneity. But there seems to be a new trend for players to rehearse what they will do. This goes back to Roger Milla of Cameroon's exotic dances with corner flags during the 1990 World Cup in Italy. Notable recent examples include the Klinsmann dive and non-league

Aylesbury Town's duck procession. Many of these pre-planned celebrations are fun and add considerably to supporters' enjoyment of the moment. But the police do feel that certain of these celebrations can cause crowd safety and order problems, which is clearly not acceptable.

The police are concerned about players who show aggressive dissent to decisions given by match officials. Television carries frequent pictures of players ranting at referees, their faces distorted with hatred; you do not need to be an expert lip reader to work out what they are shouting. The laws of football provide that dissent should be punished with a yellow card and foul or abusive language with a red, but referees at all levels of the game quickly learn to develop both a thick skin and deaf ears in the interests of ending the match with a reasonable number of players still on the pitch.

Player aggression can be directed not only at officials. In the incident involving Eric Cantona, even a spectator was on the receiving end. Fouls and misconduct towards other players are dealt with in the laws of the game. Free kicks, cautions and sendings-off generally follow. Gordon Taylor's letter drew attention to incidents where tempers have flared and there have been mass confrontations between players, making it extremely difficult for referees to control games. More seriously, from the PFA point of view, animosity between players has appeared to result in a number of cases of black players being racially abused by white players, as in allegations made by Ian Wright against Peter Schmeichel. Taylor points out these instances run contrary to football's anti-racism campaign and to what the PFA consider to be a successful integration of black, white and foreign players into the game.

Finally and, according to some reflecting the increased Europeanisation of British football, there are concerns that players are trying to con referees to gain free kicks and feigning injury to encourage booking of opposing players. The apparent dive resulting in the penalty which eliminated Leicester City from the FA Cup provides a good illustration.

The Risks

Goal celebrations which involve players running towards or even into the crowd pose three main risks to safety and order. First and most important, whenever players run off the field of play and approach the spectator accommodation, home or away, there is a serious risk of crushing injuries as excited supporters crowd forward to join in the celebrations. Second, over-celebrating in front of the opposition supporters can inflame the fans and cause disorder. Third, such player behaviour causes normally law-abiding citizens to break the law by encroaching onto the pitch, which is a criminal offence. A police spokesman told me that many of those arrested for pitch encroachment last season had been incited by over-exuberant goal celebrations.

The police feel that excess aggression, in dissention, foul play, mass confrontations or racial abuse, has an adverse effect on the atmosphere and can encourage the crowd to become more aggressive. An FA Premier League working party reported in March that a good atmosphere supports the players and is part of what the fans pay for. Any return to a more aggressive and intimidating atmosphere could have serious implications for football's commercial appeal. There is general consensus that the resurgence of British football as a successful business would be severely undermined by the return of widespread disorderly behaviour or further incidents of multiple injuries or fatalities involving spectators.

As Taylor pointed out to his delegates, excessive aggression can result in video evidence being sent to Lancaster Gate, with the result that clubs and players can face heavy financial penalties. More generally, violent player behaviour can attract the attention of both the civil and criminal courts. The common law principle set out by Lord Justice Aitken in 1932 requires that you must take reasonable care to avoid acts or omissions which you can reasonably foresee would be likely to injure your neighbour. A number of ex-players, including Paul Elliott and John Uzzell, have sought to recover damages from the neighbours who allegedly ended their careers, while Duncan Ferguson went to prison for an assault committed on the field of play.

What is Being Done?

The PFA letter seems to have raised the profile of this issue among the authorities. In April, FSOA chairman Leon Blackburn wrote both to Taylor and to John Barnwell as the League Managers Association, to emphasise the importance of players not over-celebrating goals in a dangerous manner and to invite players to discuss this concern with their respective club safety officers. And as a result of the discussions at the ACPO end of season conferences, the police have written to all the relevant authorities including the PFA. Their letter acknowledges the pressures which professional footballers face nowadays, but asks that they be mindful of the effect which their behaviour has on spectators and encourages the players to channel their exuberance and aggression into the match itself.

This correspondence is all very well, but the key issue in reducing the risks is the extent to which the players will take notice of the concerns being expressed. As Taylor stated at the ACPO conference, his members are individuals and will react differently to the enormous pressures which the commercial requirements for club success impose upon them.

While it may be almost impossible to prevent those risks which arise from spontaneous behaviour, celebratory or aggressive, it does seem reasonable to expect highly paid professional players not to pre-plan anything which might prejudice the safety, security or enjoyment of those who pay to come and watch them. Where unacceptable player behaviour does occur, the match officials, the club and the police all have a responsibility to intervene to prevent an incident escalating. Perhaps the time has come for player misbehaviour to be added to the list of key risks for which contingency plans need to be prepared.

4.5 – COMMENTARY BY JIM CHALMERS – PLAYER MISBEHAVIOUR AFFECTING CROWD SAFETY

In this 1997 article, Frosdick examines player behaviour on the pitch and the effect this can have on crowd behaviour and control. He examines how the risks to both players and spectators and questions whether more should be done to control player misconduct.

I would have welcomed Frosdick bringing club managers and the match day officials into the equation since in my experience their conduct and decisions can have an equally adverse effect on spectator behaviour. In the early days of the Football Licensing Authority it was suggested that part of the FLA Inspector's role should be to monitor pitch side incidents involving managers, players and officials which had an impact on spectator safety. This was ruled out as being beyond their statutory remit. Nevertheless it demonstrates that in the early 1990's this regulatory body was concerned that pitch side incidents did have a knock on effect on spectator safety.

Eight years after the article was written it might appear that nothing has changed – and if anything the situation has worsened. There is rarely a televised match today which does not at some stage have a player or manager screaming vitriolic abuse at the referee – and you do not need to be an expert lip reader to understand what is being said. Squabbles and fights between players are not a rare occurrence and sometimes I question whether I am watching a sport or an occasion for pugilists to display their talents. In what is an already highly charged atmosphere all it takes is the spark on the pitch to ignite the fire of spectator misbehaviour, which stewards and the police, if present, have to contend with. Any spectator reaction can and does then have an impact on the overall safety of the viewing areas.

There is a myth that a footballer on the pitch is above the law. It takes a brave police officer or club safety officer to take this one on but it has happened. As an FLA Inspector, I attended one Premier League match where the first half of the game was memorable for the numerous player incidents, which inflamed both sets of fans. The fans constantly surged to the front of the stands at every opportunity and had to be physically restrained from going onto the pitch by stewards and police. The police Chief Superintendent and club safety officer were concerned that a near riot situation was on their hands, purely because of events on the pitch, which the referee seemed incapable of controlling. At half time the police commander spoke to both managers, players and the match officials. He left them in no doubt that, if the second half continued in the same vein, then one or more players would be arrested for conduct likely to cause a breach of the peace. His comments were not well received but it had the desired effect with the second half being relatively incident free. The effect on both sets of fans was remarkable, with what in the first half had been a recipe for riot now reduced to the normal vociferous support of their teams. A real case study, which demonstrated only too clearly that pitch side incidents do have an effect on both crowd behaviour and safety.

The football authorities have long recognised this problem. In the mid 1990s, the Football League produced a video of player incidents, which had resulted in a crowd reaction, for clubs and the police to show to managers and players to explain how their conduct did affect spectator safety. The crowd reactions included surges, pitch incursions, missile throwing and attempts to assault players. At most clubs there is pre-season briefing by the club safety officer to the manager and players reminding them of their responsibilities and how any extreme action on their part will result in a reaction in the stands.

In a modern all seated stadium arguably the greatest risk to both players and spectators is when a player runs into the crowd after scoring a goal. This has the consequence of a surge to the front which could result in injury to both the fans and the player. Recently in Europe, a player lost a finger doing just this when his hand became trapped in the pitch side fencing. The hazards to players are obvious but I wonder how many clubs have conducted a health and safety risk assessment on the level of harm likely to be sustained by either a player or supporter by this celebratory behaviour.

Frosdick examines whether the consequences of player behaviour should be included as part of the event contingency planning. A contingency plan can be described as 'planning for the uncertainty of an occurrence'. Once twenty-two players, their managers and officials enter the field of play the one certainty that can be planned for is that anything could happen. It therefore makes sense for safety officers to consider the hazards which could emanate from player behaviour and to incorporate these in the event risk assessment process, as we do at Kidderminster Harriers FC.

I agree with Frosdick when he says that match officials and clubs have a responsibility to intervene in unacceptable player behaviour. However, I disagree with his view that the police should intervene. It is principally up to professional players to act professionally and when they do not for the football authorities to come down hard on them. However we regularly see on our television screens examples of player behaviour which would not be out of place in a city centre when the pubs and clubs empty onto the streets. Yet the football authorities seem incapable of controlling this.

I would argue that 'yellow' and 'red' cards do not work. Perhaps the time is right to introduce 'sin bins' where a player can spend up to fifteen minutes to reflect on his behaviour, particularly the advantage this will give to the opposing team. Unless there is more positive action taken then re-examination of this problem in a further eight years will probably show that nothing has changed. Let us hope that in football we do not yet again have to learn a lesson the hard way, as has happened in the past, with spectator deaths or serious injury being the only catalyst for change.

PART V – STEWARDING

Introduction

Part V comprises five articles which cover the stewarding of sports grounds, particularly the football stadium. The first three articles outline early research into stewarding guidelines and then explain how the training of stewards has been developed. The next article debates the evolution of high profile stewarding supported by lower profile policing. Part V then concludes with a case study which examines the death of a steward at a football ground and the legal consequences for the Club.

5.1 – Stewarding Guidelines

In 1994, Frosdick was commissioned by the British Security Industry Association to research best practice in private security company involvement in the stewarding of sports grounds. The research resulted in him producing a set of guidelines which were very much ahead of their time and which covered the surveying, planning and operation of stewarding duties. Comment is made on how this early vision for the future of stewarding has become accepted, endorsed and acted upon by the football industry in particular. Although the article was written eleven years ago the research conclusions remain equally valid in 2005.

5.2 and 5.3 – Stewards' Training

The two articles written by Frosdick and Sidney in 1996 and by Frosdick and Vaughan in 2003 examine the development of stewards' training at football grounds in England and Wales from 1991 to the present day. Both Articles provide a useful historical narrative of how stewards' training came from nothing to a professional multi-media training package. The 2003 article outlines in detail the specific training for stewards in racial awareness. Comment is made about the additional conflict management training for stewards introduced in 2005 together with the introduction of a new qualification for stewards also introduced that year.

5.4 – Higher Profile Stewarding

In this 2001 article, Frosdick examines the evolution of the principle of 'high profile stewarding supported by low profile policing' following the 1989 Hillsborough stadium disaster. Comment is made on how this has been achieved in conjunction with a whole raft of additional measures which made the principle possible.

5.5 – Accidental Death of a Steward

The article written by Frosdick and Rankin in 2001 examines the circumstances which led to the tragic death of a steward whilst performing duty at Highfield Road Stadium in Coventry, and the prosecution of the club for alleged breaches of Health and Safety legislation. Comment is made on the valuable lessons which all sports grounds can learn from this tragedy – in particular how such incidents can be prevented; and how individuals and organisations can be protected from any criminal or civil litigation consequences.

Figure 5.0.1 Eric Hope (centre), Safety Officer at KitKat (Bootham) Crescent stadium, home of York City FC, addresses assembled stewards employed by York City and private crowd control company Showsec, before a game.

5.1 – GUIDELINES FOR STEWARDING

The original citation for this article is: Frosdick, S. (1994) 'New Guidelines for Stewarding', The Groundsman, Volume 47 Number 7, July 1994, pp. 25-6.

Prior to 1989, a total of 135 people had been killed and hundreds more injured in disasters at British football grounds. As a result, no less than eight official reports had been commissioned into safety and order in football stadia. Tragically however, their recommendations and the lessons of history had too often been ignored.

Following the fatal crushing of 95 Liverpool supporters early in the FA Cup Semi-Final at the Hillsborough Stadium in Sheffield on 15th April 1989, the reports produced by Lord Justice Taylor Inquiry into the disaster proved to be the catalyst for radical change in the stadia industry.

Among the many changes, and in the absence of proper stewarding arrangements at many clubs, the police service underwent a dramatic change of role. The previous emphasis on the preservation of public order, achieved through an increasingly repressive policing style, gave way to a focus on the safety of the public as the primary policing objective. At the same time, the police service began to raise concerns about public safety in other scenarios, including pop concerts and the New Year celebrations in Trafalgar Square. Most recently, the Metropolitan Police Service engaged consultants to examine the public safety issues at the Notting Hill Carnival.

Subsequent to Hillsborough and the Taylor Reports, detailed guidance on managing public safety at diverse events in a wide variety of venues has been offered in a number of separate documents. These have included publications by the Home Office, the Home Affairs Committee of the House of Commons, Health and Safety Commission, Health and Safety Executive, Football Licensing Authority, Football Stadia Advisory Design Council, the Football League Ltd, Institution of Structural Engineers, Association of County Councils and the Sound and Communication Industries Federation.

A report originating from Greater Manchester Police and the recent scrutiny into the public order commitments of the Metropolitan Police Service show that the current trend in public safety at stadia and sporting venues is for the police to seek to reduce their manpower commitments and to concentrate their resources on public order and emergency management functions. At the same time, venue management have begun both to assume their proper responsibilities for ensuring the safety of their customers and also to seek to reduce the escalating costs of special services of police.

The Football Licensing Authority have begun monitoring the quality of stewarding schemes and have been urged by Parliament not to hesitate to use their powers against clubs where stewarding standards are unacceptable.

Venue management are increasingly requiring security services for stewarding, cash handling, revenue protection (e.g. pirate merchandising and ticket touting), static guarding and access control.

The assumption by stewards of functions previously undertaken by police, such as searching spectators and enforcing ground regulations, has been undertaken in a variety of ways: in house development through the employment of retired police officers as safety managers; in house development with guidance from external consultants from the private security industry; mixed in house and agency stewards, deployed either in mixed teams or for separate functions; or complete contracting out of the security services required. The latter three all involve an input from the private security industry. Thus within the football industry and elsewhere, there is a growing need for specialist advice and services and it seems likely that venue management may increasingly look to the private sector to provide these.

The commercial possibilities offered by this move to higher profile stewarding, coupled with the wish to provide guidelines and standards for the industry, persuaded the British Security Industry Association to make research funds available to the Centre for Risk Management at Cranfield University. Subsequently, as an ancillary to my own MSc research into the strategic management of public safety risks in stadia and sporting venues, I produced a set of guidelines for the stadia stewarding industry.

These Guidelines for the Surveying, Planning and Operation of Stewarding Services in Stadia and Sporting Venues have now been published by the British Security Industry Association. They are intended to assist specifiers, users and providers of stewarding and other security services in selecting the level of services best suited to the particular risks at the stadium or other sporting venue concerned and the requirements of any certifying authority.

The guidelines are primarily intended to apply to sports grounds designated under the Safety at Sports Grounds Act 1975 and other venues incorporating stands designated under the Fire Safety and Safety of Places of Sport Act 1987. All such designated grounds and stands require safety certification by the appropriate local authority. The provisions may however be equally applicable to other stadia and sporting venues, for example a non-league football club hosting major opposition in the FA Cup, and other events such as a concert being held under the authority of a local authority music and dancing licence.

The main objectives of the guidelines are as follows: to establish and maintain standards of stewarding services by private contractors in stadia and sporting venues; to assist specifiers and users to determine the appropriate level of service required for given circumstances; to assist providers of stewarding services to meet specifier or user requirements; to promulgate best practice; and to define terms used within the stewarding services industry.

In addition to pulling together into one document the many recommendations made in previous publications, account has also been taken of best practice found at various stadia and sporting venues in the United Kingdom. Strategies, operational tactics and the extent of partnership between the various agencies involved vary considerably from ground to ground. There are many alternative methods of achieving good spectator safety. Most importantly therefore, the guidelines seek to acknowledge the diversity of practice and culture to be found and do not advocate any *one* best way.

The main emphasis is placed throughout on the surveying, planning and operation of stewarding services. Whilst other security services may be required and are therefore acknowledged, references to them are less extensive.

In respect of surveying and planning, advice is given on control rooms, general venue design and, particularly, on liaison and co-operation with other agencies for risk identification and emergency planning purposes. A checklist is provided of the more commonly recognised risks, from lock out to bomb scare to floodlight failure. A range of issues to be considered when determining staffing levels is also included. For staffing in general, guidelines are included on staff selection, vetting and training. There is particular emphasis on training, both in general and for specialist posts, and also on staff assessment. Reference is also made to the need for external validation of training and assessment.

In terms of operations, advice is given on site supervision, briefing and debriefing, and equipment and facilities for stewarding staff. The basic duties of stewards are set out, and details provided of the various different types of specialist stewarding duties, from monitoring signs of crowd distress to fire safety patrols. Finally, guidelines are also provided on the different types of documentation and record keeping required in support of the planning and operating functions.

In the view of the Home Affairs Committee of the House of Commons and many others, higher profile stewarding supported by lower profile policing is the way forward for the stadia safety industry. Thus the quality of stewarding, which should be considered by police commanders when determining the numbers and costs of police to be deployed, is becoming an ever more important issue.

These Guidelines for the Surveying, Planning and Operation of Stewarding Services in Stadia and Sporting Venues should therefore be of interest to various parties involved in the management of sporting events, including: Police Commanders seeking to encourage improvements in stewarding quality so as to be able to reduce their own manpower commitments; stadium managers and safety officers looking both to assume their proper responsibilities and to cut their policing costs; local certifying authority officers seeking standards against which to measure the quality of the stewarding operation at stadia under their jurisdiction; and private security companies interested in entering the market to provide this type of service.

5.1 – COMMENTARY BY JIM CHALMERS – STEWARDING GUIDELINES

The article written by Frosdick in 1994 was the result of his research on behalf of the British Security Industry Association (BSIA) in 1993 when he examined the opportunities for the private security industry in the stewarding of sports grounds. At the time of his research Frosdick was a Police Inspector with the Metropolitan Police Service. It is perhaps regrettable that his research had to be funded by the private security industry since the guidelines he produced related more to the safe stewarding of sports grounds than to the more traditional role of a security officer who might also be employed as a safety steward at a sporting event. However at the time the guidelines were written there was still confusion and debate about what had primacy at a sports ground – was it spectator safety or ground security?

In 1994, the appointment of professional safety officers was only just beginning. As the police refocused their role and gradually withdrew from inside sports grounds, the introduction of well-trained and equipped stewards was also in its infancy. Police control rooms were being replaced by stadium control rooms. CCTV was beginning to be used for ground safety instead of just police surveillance. Such changes were all part of the developments examined by Frosdick. It is regrettable that his vision for the future did not interface at the time with the work being done by the joint football authorities safety management working party, established in 1993, since they were examining the same issues raised by Frosdick in his research. As a member of that working party, I can see that much of what Frosdick produced in his stewarding guidelines was very much reflected in the future stewarding guidance which would be issued by the football authorities.

Much of what Frosdick suggested can be seen in evidence today. This is why the guidelines are in many ways remarkable, since they were produced well ahead of stewarding guidance published by either the football industry or the sports regulatory bodies. As an example his guidelines on steward selection, training and assessment were reflected in the football authorities 1995 publication 'Stewarding and Safety Management at Football Grounds'. They can also be found in the 1997 edition of the 'Green Guide' to Safety at Sports Grounds. What we now accept in 2005 as the norm for stewards training and assessment were key issues identified by Frosdick in 1994.

His vision for the future in terms of surveying and planning has also stood the test of time. Much has been published about safety systems and stadium design and comparison between his guidelines and the most recent edition of the 'Green Guide' again shows how far ahead he was in his vision for the future of sports grounds. His comments on the importance of contingency planning and briefing/de-briefing were later reflected in guidance on both subjects issued by the Football Licensing Authority in 1997 (and updated in 2002 and 2004). It is only by comparing his original guidelines with relevant guidance subsequently issued by others that his remarkable vision can truly be judged.

The 1994 article does have a parallel with the Private Security Industry Act 2001 and the authority this gave for the establishment of the Security Industry Authority in 2003. These new controls were introduced to regulate door supervisors, wheel clampers and manned guards, who previously had been totally unregulated. I would suggest that in 1994 Frosdick could see that the private security industry recognised the commercial opportunities presented by providing security personnel for stewarding purposes at sporting events. Frosdick foresaw the risks of an uncontrolled and unregulated influx of security personnel into the football safety industry and his guidelines sought to address this. In this context his guidelines remain valid today, albeit that much of what he produced has since been absorbed into other guidance, most notably the fourth edition of the 'Green Guide'.

Frosdick's involvement in the development of stewarding at sporting events, particularly football, continues to the present day, as articles 5.2 and 5.3 will illustrate. His 1994 guidelines, which remained an official BSIA publication until June 2001, were very much an embryo for the growth of the massive safety management and stewarding industry which we have in sport in the present day.

Figure 5.1.1 Showsec stewards accompany officials off the pitch at the end of a game at KitKat (Bootham) Crescent stadium, home of York City FC.

5.2 – STEWARDS TRAINING GOES MULTI-MEDIA

The original citation for this article is: Frosdick, S. and Sidney, J. (1996) 'Stewards Training Goes Multi-Media', *Football Management*, Volume 4 Issue 3, Summer 1996, p 26.

Football Safety Officers' Association members Steve Frosdick and John Sidney report on the new training package for stewarding at football grounds.

Throughout the 1970s and 1980s, football matchdays often resembled military operations. Huge numbers of police were employed on tactics which achieved control, but at the expense of safety and comfort. Subsequent to the Hillsborough tragedy, the Home Affairs Committee of the House of Commons investigated the policing of football hooliganism. Their 1991 report described 'higher profile stewarding, supported by low profile policing' as the way forward.

And great strides have been taken towards that vision. At Nottingham Forest, for example, a typical match in 1989 saw over 150 police officers inside the ground, supported by about 75 stewards. Compare this with the season just ended, when most Forest matches were policed by 250 or more stewards, supported by just 22 police officers. And several clubs, including some in the Premiership, have managed to police selected matches with no police whatsoever in the ground. Professional stewarding has started to become a reality.

This role reversal has been driven by three key factors. First were the concerns raised both by the authorities and by practitioners, that standards varied so greatly from ground to ground. Second was a desire to demonstrate the ability to self-regulate rather than be dictated to. Third was the need to achieve good quality training with value for money.

But what has actually been done to bring about such change? In response to the concerns raised, the Football League published their 'Guide to the Appointment, Training and Duties of Football League Club Stewards' in July 1991. This guide, supported by a video, included five outline training modules and provided a useful first set of recommendations for the industry. Subsequent changes in the stadium environment, in policing policy and in Football Trust grant aid for stewarding, as well as the formation of the Football Safety Officers' Association (FSOA), resulted in a meeting in October 1993, at which a working party was set up to update the Football League publication. As a result of that work, a set of comprehensive guidelines on 'Stewarding and Safety Management at Football Grounds' were published by the football authorities early in 1995. The guidelines recommended six modules for stewards training and set out the aims and objectives of each module, together with the performance criteria the training was intended to achieve. The six modules were:

- General Responsibilities;
- Maintenance of a Safe Environment;
- Response to Spectators;
- Emergency Aid;
- Basic Fire Safety Awareness; and
- Contingency and Evacuation Plan Training.

The six modules provided an excellent framework, and at the FSOA annual general meeting in March 1995, both practitioners and the authorities agreed to take things a step further. A second working party was formed with the brief to develop a package of training materials to put some flesh on the bones of the six modules. This venture was now a partnership involving the Football League, the Football Association, the FA Premier League and the Football Safety Officers' Association, and was undertaken in consultation with the Football Licensing Authority.

In line with the second and third of the key factors outlined above, the working party decided against the level two NVQ in crowd management, which they considered would be too expensive for the clubs and unpopular with the stewards. It was therefore decided to develop the training package using in-house expertise and the idea of a multi-media, computerised package quickly won support. After presentations from several computer companies, the working party decided the package should be prepared using Microsoft Powerpoint electronic presentation software. This approach was endorsed by the FSOA September conference and a consultant was appointed to undertake the necessary work.

By Christmas, the working party had edited the first draft and a second draft was then circulated to the FSOA membership. The various reviewers' comments were collated and a third draft was presented to the FSOA annual general meeting in March 1996. The draft package was debated, approved and the final product commissioned.

The new multi-media package was launched by the Football League President, Gordon McKeag, at a national seminar at West Bromwich Albion on 5th June 1996. The final package was demonstrated and copies were issued to the representatives of each club. The launch also included a useful session on good practice for trainers in giving presentations.

The package is made up of the six modules set out in the original framework, together with an initial familiarisation session and an induction training module for newly appointed stewards. The package comes in a presentation binder and has three main components: computer disks containing electronic full-colour presentations; black and white slide masters for use in producing acetates; and notes pages to help the trainer.

Using the Microsoft Powerpoint Viewer, the electronic presentations can be used in several different ways. If training a large number of stewards, trainers can hire a liquid crystal display (LCD) tablet and project the presentations onto a large screen using a high intensity overhead projector. This is a very effective method and gives a very professional impression to the audience. With a smaller group, trainers can sit their stewards round an ordinary computer screen. Alternatively, stewards can be allowed to teach themselves by sitting in front of the computer and operating the presentation alone.

Trainers who have their own Microsoft Powerpoint software can also edit the package. They can insert clipart, change the colours or rewrite some of the text. But even without a computer, trainers can still show all the slides using an ordinary overhead projector. The package includes a black and white master of every single slide. Trainers can decide which slides they want to reproduce, and simply photocopy them onto acetate.

The modules are made up of topics, each of which is covered by a notes page, containing:

- the relevant slide;

- the instructional objective;

- the suggested instructional content;

- some topics, for example football legislation, are similar for all stadia and in these cases, additional slides and notes are provided for the trainer's use;

- other topics, for example standards of dress and appearance, are different for each stadium. In these cases, the content will have to be determined locally, although a suitable title slide with clipart is usually provided.

- the assessment criteria, i.e. what the steward is expected to have learned as a result of receiving the training; and

- the suggested assessment method(s).

The introductory modules cover a wide range of topics, including the broad framework of safety management, stadium layout, legislation, ground regulations, stewards duties and conduct. The next module deals with crowd behaviour, crowd control, stadium safety features and equipment. The following module addresses spectator breaches of safety as well as general customer care issues. The next two modules provide the basic knowledge needed to cope with a medical or fire situation pending the arrival of the emergency services, whilst the final module covers contingency plans, safe evacuation and the accompanying practical exercises.

The football authorities recognise that it will take some time for all stewards to be fully trained and it is certainly not intended that all stewards should be trained in every aspect of the package before they are deployed. The timescales recommended are as follows:

- Before any deployment whatsoever – an initial familiarisation session;
- A minimum of four matches in company with a qualified steward;
- Before being allowed to work alone – induction training module;
- Within twelve months of Induction Training – completion of all six modules, designation as a qualified steward and award of a certificate;
- Within three years – completion of all six modules by existing qualified stewards.

The football authorities have recommended the package as a best practice training aid which represents the minimum requirement for the training of stewards at football grounds. The package is a resource on which trainers should be able to draw in preparing their own courses. They do not have to follow it letter for letter. If they already have their own way of meeting a topic objective, they are free to carry on using it. But the early feedback is encouraging and it seems likely that most trainers will be happy to use the slides and notes pages as the basis for their training delivery. The work, however, continues and the working party are keen to move forward with the next rungs on the training ladder, including assessment and the training of supervisors and safety managers. Higher profile and more professional stewarding remains the way ahead.

5.3 – DEALING WITH RACISM: TRAINING STEWARDS

The original citation for this article is: Frosdick, S. and Vaughan, A. (2003) 'Racism in Retreat', *Stadium & Arena Management*, October 2003, pp. 29-31.

Steve Frosdick and Alison Vaughan explain how stewards in the UK are now receiving special training on how to deal with racism in football.

Throughout the European football family, everyone is agreed that racism in football is a problem which needs to be dealt with, and it is welcome that UEFA are leading on a range of anti-racist measures under their 'unite against racism' initiative. But of course these various strategies and polices need to be translated into action – by stadium management and, above all, by the stewards. This article therefore focuses on the work recently undertaken to develop an anti-racism training module within the UK 'Training Package for Stewarding at Football Grounds'.

The original training package dates from 1996 and covered six training modules: general responsibilities; maintaining a safe environment; response to spectators, emergency aid, basic fire safety; and contingency and evacuation plans. Due to the passage of time, it was decided to commission an update in late 2002. And as a result of recommendations made in two UK Government enquiries, the need was identified for the revised package to include training module on how to deal with racism. The updating work was undertaken in close consultation with 'Kick It Out', a UK organisation which campaigns to kick racism out of football.

The revised training package was launched in March 2003 and now includes a seventh module on dealing with racism and disability discrimination. This new training module aims to give stewards the necessary understanding and confidence to deal with racism in football grounds. It also seeks to ensure that stewards respond appropriately to racist incidents.

The 'dealing with racism' training covers three main topics. 'What is racism?' gives an overview of the problem, why it needs to be addressed, the current law and what constitutes racially offensive language and behaviour. The second topic looks at the action stewards can and should take. The final topic considers the implications of any action and provides a summary of the training.

Like the rest of the training package, the new training module has been written using Microsoft PowerPoint presentation software. So, for each topic, there are slides and notes pages which include the learning objectives, the suggested training content, the assessment criteria, and the suggested assessment methods. In fact the full training package of seven modules contains 300 slides and notes pages.

So what is racism? By the end of this topic, the intention is that the steward will: understand what racism is; appreciate why racism is a problem in football; be aware of the current law on racist chanting and abuse; understand what constitutes racially offensive language and behaviour; and appreciate the importance of their role in dealing with racism in football.

We would begin this topic by asking the stewards to suggest what racism is. We might want to write down their answers on a board or flipchart. We would then give two examples as a basis for discussion. First that racism is the belief that people are inferior because they are a different colour, or from a different country or part of the world or have different religious beliefs. Second that racism is a prejudice or judgement based solely on ignorance.

We would point out that, for many years, football has been associated with racism and racist abuse. During the 1970s and 1980s, monkey chanting and banana throwing were a regular sight during matches. Well known black players spoke out about the constant abuse they received during matches. Many black players actually left the game because they could (and would) not accept the abuse that they were continually subjected to.

But racism is still a major problem in football. Black players still receive abuse during matches. For many fans, the match day experience is stilled soured by racial abuse and even violence. We could read out recent letters from fans complaining about racist behaviour at our own club. We would emphasise the low number of ethnic minority fans that attend matches, which on average is less than 1% in the UK.

Racist abuse within grounds is very clearly against the law. The UK law states that it is, 'An offence to engage or take part in chanting of an (indecent or) racialist nature at a designated football match'. Chanting is defined as the repeated uttering of words. Whilst chanting was originally only an offence if there was more than one perpetrator, the law has been amended to state that the offence is committed 'whether alone or in concert with one or more others'. The London Metropolitan Police recently achieved a successful prosecution on the basis of two different comments by one individual.

The law also defines what is meant by 'racialist nature': namely, 'Anything that is threatening or abusive or insulting to a person by reason of colour, race, nationality or ethnic or national origins'. So individuals and groups who repeatedly utter racist abuse are clearly committing a criminal offence.

There is often confusion by stewards – and the public in general – surrounding words and terms that are deemed as racist and might be seen as insulting or racially offensive. This confusion is not helped by the fact that the use of words may change over time and may depend on where you live. So we would use some examples to facilitate a discussion about which words or expressions are racist and racially offensive and which are not. In many European countries, there is increased sensitivity about use of inappropriate language and it is not now acceptable to use many terms that might have been freely used in the 1970s and 80s.

We anticipate that stewards may ask about other kinds of insults and abuse. Stewards do encounter all sorts of bad language and behaviour in the stadium – all of which may need to be dealt with – but we would emphasise that the focus of this training is on very negative experiences suffered by Black, Asian and other visible ethnic minority players and fans. So we would emphasise why the role of stewards in recognising and tackling racism within the ground is so important.

Stewards have the responsibility to ensure that anyone watching a football match can do so in a safe and pleasant environment. For many this will mean without the fear of being subjected to or witnessing racist abuse, whether this is directed at a fan or at a player. For most fans and spectators, match day stewards are their first point of contact within the ground, especially if any problem should arise. Stewards are often seen as ambassadors for their clubs, therefore how an individual steward deals with an incident of racism is a reflection of how the club overall is perceived to be dealing with the problem.

The next topic looks more specifically at the expectations football has of its stewards and at the actions that they are expected to take to deal with racism. By the end of this topic, we expect that the steward will: clearly understand that they are expected to respond to racist incidents and not to ignore them; and know, understand and be able to apply the a simple four step process for dealing with racist incidents.

We would begin by outlining the expectations that are placed on match day stewards. In terms of dealing with racist abuse these include expectations from fans, players, the football club, the police, the football industry as a whole, the media, campaigners and community groups as well as local communities. We would use a small groups exercise to get stewards thinking about the expectations that are placed upon them. We would give each of the different groups one of the categories and ask them to discuss what they believe the expectations to be. The small group discussions would be fed back to the whole group. Once the whole group has discussed the expectations we would use a pre-recorded message from a famous player or senior club representative to reinforce the expectations that stewards will take appropriate action to deal with racist incidents.

In 1998, a UK government report called for a simple standard procedure or action plan for dealing with racist incidents at all football grounds. 'Kick It Out', with the support of the football authorities, has identified four main elements in this standard process. These are to listen, to assess the situation, to take the appropriate action, and to make a suitable report. The process is always carried out in the same order.

A steward is likely to be made aware of racist behaviour in one of two ways: either they hear it personally, or it is reported to them by another fan. Stewards should listen to individual and group chants as part of their general monitoring of the crowd. However, it is accepted that in large crowds it is often difficult to hear individual comments and chants. If racism is reported to them, stewards should deal with the person complaining calmly, courteously and sensitively. They should listen carefully to establish the precise nature of the complaint – people complain to stewards about racist comments they have heard because they are offended by them, even if they are not the direct recipients of the abuse. Stewards should acknowledge the complaint – clubs are working hard to encourage people who have witnessed racism to complain about it, rather than to ignore it. Therefore stewards should always acknowledge the complaint and on no account ignore the problem and turn a deaf ear.

The next stage of the process is for stewards to make an assessment of the situation. What was actually said? How many times was it said? How many people are involved? What are the risks involved?

Risks can include risks to the complainant (if there is one), to the crowd, to the steward and to the offender. As always, the safety of all those involved will be the paramount consideration.

Points to consider will include, for example, is it safe for the steward to enter the crowd situation alone or is back up required? Will a quiet word of advice to the offender suffice or is firmer action needed? Are there any standing instructions for dealing with incidents of this type? Should the steward consult with a supervisor or the Control Room before deciding what to do?

The assessment will determine the third stage of the process, which is to decide what action should be taken. Action does not necessarily involve pulling someone out of the crowd and having them arrested, it can mean any number of things. For example, the steward's first course of action may simply be to report the incident to a supervisor because it may not be appropriate to deal with it on their own, or safe to respond in the given environment. The action may be to check the ticket (if there is one) and take a note of the offender's seat number or terrace location. It may be to give some sort of advice or warning. It may be to call for help and consider ejecting the offender from the ground. Some cases may justify an arrest being made, either by calling for police assistance or even by the stewards themselves.

Finally, the steward should always make a suitable report of what has happened. This may simply be a verbal report to a supervisor, either at the time or when debriefing at the end of the match. However, in many cases, a written report will be needed. It is important to keep an accurate written record of incidents, even if the steward takes no physical action, or even if the offender is simply warned and does not then re-offend. Written reports will often take the form of an incident card, which is submitted through the supervisor as part of the debriefing process. It is good practice to encourage stewards to complete such incident cards, which can then be filed for future reference in the event of a query, complaint or other legal action.

The final topic is designed both to reinforce some of the earlier messages and to give stewards the confidence to be able and willing to deal with racist incidents in football grounds.

The summary should include the following points:

- Racism and racist incidents do still exist within the game.
- It is the responsibility of all those involved in football to rid the game of this problem.
- Stewards have a vital role to play in eradicating the problem.
- Racism spoils the match day experience for many people, whether they are the intended victims or not.
- All fans should be able to watch the game in a pleasant and safe environment.
- Every week people make complaints to `kick it out' about racist incidents that have occurred.
- Complaints are made because the large majority of fans do not want their clubs associated with racism.
- It is essential that stewards meet these expectations and do not ignore the problem.

It may be useful to reassure stewards that dealing with the problem of racism can often seem a daunting task, but the process established (listen, assess, act and report) should ensure that the problem is dealt with effectively. Stewards can – and do – make a positive difference to the match day experience of many fans each week, up and down the country.

As they already are with the other football offences of pitch incursion and missile throwing, stewards should be prepared to support the criminal justice process by making a witness statement to the police and, if necessary, giving evidence as a witness in a criminal trial. The club and the police must in turn promise to support stewards who are prepared to make a positive difference in this way.

We would then show a short pre-recorded message of encouragement from Piara Powar, the National Co-ordinator of 'Kick It Out'. Having thus encouraged the stewards to be willing to take action, we would facilitate a discussion, asking the stewards to decide what form of action would be the most appropriate in each of the four situations:

- A complaint about one fan making a single offensive comment;
- Hearing one fan shouting a single offensive comment to a player;
- Two fans continually shouting racist comments at opposing fans;
- A complaint that a large group of fans are continually chanting racist abuse at a player.

We would then facilitate a second discussion about four examples of incidents and the action taken by stewards. We would ask the stewards to decide what type of report should be made and who is responsible for making it.

- Offensive comment heard but not repeated – seat number taken;
- Several offensive comments by two fans – warned of consequences if repeated;
- Repeated abuse by one fan – warned but re-offends – assistance sought – ejected from ground;
- Whole area singing racially offensive songs – Supervisor informed.

There are no right answers for these discussion sessions, but we believe that the overall effect will be to give the stewards confidence to take the appropriate action and make a suitable report.

To conclude, we hope that our new training module on dealing with racism has given you some food for thought. This training could be used not just for stewards, but for other private security staff or even for the police. If we're serious about making progress in kicking racism out of football, then we need to get on and do something about it.

Figure 5.3.1 Stewards employed by York City and private crowd control company Showsec await a briefing before a game.

5.2 AND 5.3 – COMMENTARY BY JIM CHALMERS – STEWARDS TRAINING

Article 5.2, written by Frosdick and Sidney in 1996, outlines the background to the introduction in 1995 of the first national training course for football stewards in England and Wales. In article 5.3 in 2003, Frosdick and Vaughan bring the training package up to date and introduce the additional training module on racism and disability discrimination. In respect of both articles I must declare a personal interest since I was a member of the joint football authorities safety management working party between 1993 and 2003. This was the group which oversaw the introduction of the training package referred to in both articles. The working party comprised representatives from the Football Association Premier League, the Football League, the Football Safety Officers' Association and I was one of two Inspectors representing the Football Licensing Authority. My interest and involvement in spectator safety management and stewarding continues to the present day as President of the Football Safety Officers' Association and as Deputy Safety Officer at Kidderminster Harriers FC, where all the training and assessment theory has to be translated into practice – as it does throughout football and in other sports.

Although the first edition of the 'Green Guide' to Safety at Sports Grounds, published in 1973, referred to the need for stewards to be trained, this meant all things to all people. This is perhaps not surprising when you consider that it was not until the fourth edition of the 'Green Guide', published in 1997, that the definition of a 'steward' first appeared. Not only did football clubs not agree on what stewards training involved but the local authorities responsible for the safety certification of football grounds similarly could not agree. Most safety certificates simply said that stewards should be 'trained', with no clarity as to what this meant.

For example in the 1970s and 1980s, as a police officer, I saw people turn up at a ground on match day and say they wanted to be a steward. Their training amounted to a senior steward giving them a jacket and saying 'stand there and if I need you I'll come and get you'. I can recall at a major Midlands club over 250 stewards reported for the first game of the season to be welcomed en masse by the Chief Steward – and this welcome amounted to the total 'training' for the season.

In the 1990s, when the football authorities introduced a stewards training certificate, over 50 stewards attended a thirty-minute talk by the safety officer on changes to the ground. As they left at the end on their way to the bar they were all handed their training qualification certificates. This then was the reality of steward training, or rather the lack of it, from the 1960s to the early 1990s.

It was not surprising therefore that when our working party was formed in 1993 there was universal agreement that things had to change if football stewards were ever to be considered as professional and competent. After nearly two years of deliberations our group produced the booklet 'Stewarding and Safety Management at Football Grounds', otherwise known as the 'Red Book'. Whilst Frosdick and Sidney concentrate in their article on the six stewards' training modules, the 'Red Book' contained much more guidance on both safety management and stewarding. This was all later incorporated in the fourth edition of the 'Green Guide'.

In article 5.2 Frosdick is very modest since the working party all agreed that the production of the training criteria was not enough in itself. The group identified a need for a training package in support of the six modules if consistency in training delivery and assessment was to be achieved. The football authorities appointed Frosdick as the project manager to produce the training package based on the 'Red Book' training criteria. This proved a wise decision with Frosdick delivering the multi-media training package on time and to the training specifications set. All Clubs accepted the package and since 1996 this has been the catalyst for a revolution in stewards training in football and in other sports in both the UK and overseas. The project proved the value of integrating the practical experience and knowledge of the working party with the academic and technical expertise of Frosdick. The training package has from the outset been a success story for the football industry in England and Wales, with the benefits extending into other sports and other countries.

However the training package only delivers the visual aids and scripts to assist the trainers. In my time as an FLA Inspector I attended numerous stewards training sessions at Premier and Football League Clubs when the content ranged from excellent to abysmal. Sadly, some safety officers still believe that being a competent safety officer is enough to make them a competent trainer. As a result, many stewards and I have had to endure some dire training sessions. The training package clearly identifies the need for the training to be delivered by competent persons but in 2005 the training is still being delivered at some clubs by inadequate trainers.

Time spent in training delivery can also vary widely. One private security company delivers all the six modules in just eight hours, whereas stewards at the majority of clubs have between eighteen and twenty four hours instruction to complete the training. It is this inconsistency which is the only weakness in the stewards training programme. Although the package was introduced in 1996 there has never been any research to examine whether the time spent in training has any correlation with spectator behaviour. In other words does any extra time spent in stewards training result in improved spectator behaviour? I think the time is long overdue for such a research project as an independent evaluation of the benefits for the football customers of time spent in stewards training.

Frosdick and Sidney conclude article 5.2 by commenting that the work of the football authorities working party would continue and the 2003 article outlines the developments in the intervening seven years. In 1998 the 'Red Book' was updated with the title 'Safety Management at Football Grounds'. The most significant development was a new emphasis on assessment, which resulted in the introduction of the 'Football Stewarding Qualification' (the FSQ) the following year.

In 2002 Frosdick was appointed project manager to update the 1996 package to reflect concerns over the adequacy of stewards training to deal with incidents of racism and disability discrimination. Article 5.3 deals with part of the new Module Seven. Frosdick and Vaughan exclude any reference to disability discrimination in their article. This is understandable since the conference paper from which the article came focused only on racism. However it is also unfortunate, since the football authorities place equal emphasis on the need for football clubs to tackle both issues in a positive and proactive way. Only including half the new module could give a false and misleading impression that disability discrimination is not being treated with the same importance as racism. In reality stewards training and assessment places an equal weight on both issues and I would have welcomed Frosdick doing the same in the article.

Having said that, the article provides a very accurate summary of the problems faced in tackling racism and how Module Seven can help overcome these. I am pleased that due recognition has been given to the contribution by Vaughan of the 'Kick It Out' campaign since this organisation has done more than most in tackling the cancer of racist behaviour at football grounds. It also recognises the importance of having competent persons involved in preparing the training materials for both racism and disability discrimination. None of the working party had such competencies and throughout the development of the training package it was always recognised that the credibility of the content relied heavily on the skills, knowledge and experience of those involved in drafting each of the modules. The updated training package was delivered on time and in the same multi-media format as the first edition.

Has Module Seven achieved the desired objectives referred to by Frosdick and Vaughan? Having been involved in the practical delivery of Module Seven at my Club I would give this an unqualified 'Yes'. The visual aids and scripted notes work well and the feedback from our stewards has been very positive. Stewards have commented how the training has helped their understanding of racism and disability issues in their everyday lives. I recall one of my stewards saying after the training that he never really understood what racism meant and how for years, subconsciously and unintentionally, his conduct could well have been considered as racist. There was similar feedback on their lack of understanding of how disability discrimination can extend beyond just wheelchair users. My stewards would certainly support the argument that the training has helped make them better citizens.

Module Seven has filled a gap in stewards training at football grounds but I would question to what extent this training has been adopted by other sports. I have heard comment from safety officers in other sports that because they have received no complaints of either racism or disability discrimination then they do not have a problem. I question the accuracy of such statements given my own experiences when stewards did not understand what either racism or disability discrimination actually meant. In my opinion the content of Module Seven should be compulsory training for anyone employed in the management and stewarding of any sporting venue. To do otherwise can only amount to complacency about customer care issues, which extend way beyond the realms of football.

However, it is not enough just to produce a superb training package; it must be delivered by persons competent to do so. I am not sure that an elderly, white, male, middle class safety officer who has never been the subject of racism or disability discrimination is qualified to speak on this subject – but many do so. The more enlightened clubs use the services of people who are involved in dealing with racism or disability in the community and whose understanding of the real life problems outside football make them much better qualified to teach the topics.

Without being complacent about racism, the overall situation is improving and when complaints are made, positive action speaks louder than words. During the 2004/05 season at my own ground we received a complaint of racist abuse directed towards a visiting black player from one of our home areas. Sadly the complaint was made by a fan as he left the ground at the end of the game and this showed a weakness in our system, with him having neither an awareness of how to report such conduct nor confidence in how we would respond to the complaint. We responded at the next home game by employing four police officers in plain clothes in the area of the terracing where the incident occurred with a brief to listen for and arrest anyone using racist language. We used police officers to provide an independent element in the complaint investigation and to provide the evidence in support of any prosecution. Since the complaint was made there has been no repetition of racist conduct but we were able to demonstrate a positive response to the complainant. I wonder whether the Spanish authorities have responded to serious incidents of racist conduct at football matches in their country during 2004/05 with the same vigour?

Since the 2003 article the development of stewards training and assessment has continued. In March 2005 a new Module Eight was introduced to provide football stewards with additional training in conflict management. The subject was already covered in another module but a specific module was considered necessary to respond to criticisms made by the Security Industries Authority. The new module deals with the methods of managing conflict and includes specific training for those 'designated' stewards who have either to arrest or eject someone from a football ground. The six main topics cover: the steward and the law; conflict resolution; initial action and options; the use of force; personal safety; and escort and safety ejection techniques.

Also in March 2005 the football authorities introduced a new qualification – the Certificate in Event and Matchday Stewarding. This can be awarded to stewards who have successfully completed their training in the eight modules and who have been assessed as competent in the workplace. This qualification will be additional to the current Level 2 NVQ in Spectator Care and Control, which many football stewards already hold, including ours at Aggborough Stadium. The certificate, which will be an accredited qualification, will replace the FSQ, which the football authorities introduced in 1999 as an alternative to the NVQ.

The FSQ has always been criticised from outside the football industry for not being a properly accredited qualification and the new certificate will address this. It will also resolve the problem of stewards who also perform duties at cricket, rugby and horse racing venues. The certificate is an event qualification as distinct from being football specific, thus making the stewards who hold the certificate qualified to perform duties at any sporting event.

It is envisaged it will take a steward between 60 and 120 hours to complete the training and assessment necessary to qualify for the award of the certificate. Stewards are part-time employees and many do not even receive the minimum hourly rate of pay recommended by government. This leads me to question just how far the level of stewards training can or should go. What started in 1996 with six modules has nine years later extended to eight with the most recent modules being more academically challenging, requiring an understanding of theory and practice in racial awareness, disability discrimination and conflict management.

At our club, with just 60 stewards to train and assess, the number is manageable with an average annual wastage of less than 10%. However for major football clubs employing between 200 and 1,000 stewards, with some having an annual wastage rate of 40%, the training and assessment processes are a major undertaking. I therefore think that the football regulatory bodies should carefully review any further training requirements, taking account of the existing heavy time demands to achieve the present levels of competency. I would suggest that, given the nature of stewarding, there is a limit to just how much information a steward can assimilate. Anyone who has either undertaken the training or sat through the existing modules will understand precisely what I mean. If the football authorities are not careful, more time will be spent in training and assessment than actually doing the job.

Stewards' training remains important, but clubs and stewards will be judged on how that training is delivered in practice. The UK football industry can take great pride in the way they have led the world in matters of stewards training and assessment. I believe other sports in the UK and other countries have a long way to go to match what has been achieved by the football industry in England and Wales. With the passage of time, there is a danger of historical amnesia clouding the memory of just how far football stewards training has been developed since the football-working group first met in 1993. I hope that these articles and this commentary ensure that present and future safety practitioners, sports regulatory authorities, police, academics and students never lose sight of that history.

5.4 – LESS POLICE: MORE STEWARDS

The original citation for this article is: Frosdick, S. (2001) 'Switch to Safety', *Stadium and Arena Management*, October 2001, pp. 9-10.

Steve Frosdick gave a paper entitled 'Lower Profile Policing and Higher Profile Stewarding: An Evaluation' at the Stadia and Arena 2001 conference.

Safety and security is a key function of stadium and arena management. In the last few months alone we have seen close to 200 spectators killed in stadium disasters in South Africa, Congo, Iran and Ghana. Crowd misbehaviour also has a long association with stadium and arena events. But safety and security needs to be located within a broader risk management framework. Managers need to strike an appropriate balance between safety and security and three other competing demands. Commercial pressures mean they must optimise the commercial viability of the venue and its events. Spectator demands for excitement and enjoyment require credible events staged in comfortable surroundings, whilst any negative effects which the venue and event may have on the outside world must be kept to a minimum.

The key to successful risk management is striking an appropriate balance between these demands. So what I want to do is to outline the various changes in policing and stewarding and then evaluate the changes from these four different perspectives.

Historically, the british police had a very negative attitude towards sports fans – particularly football supporters. Officers described fans as 'animals'. And as animals they have been treated. We've seen supporters – especially the away fans – herded like cattle from their transport to the stadium. On arrival they have been shown into a pen and then caged behind a fence. They've been carefully segregated from the home fans and then quickly herded away as soon as the match is over. Of course segregation is important – but there are ways of doing it with ticketing, access control and stewarding.

Policing has, historically, been arbitrary and unfair, with fans arrested or thrown out of the ground for no good reason. I say 'historically' but to what extent are these comments still true today. Consider the contrast in policing styles for the Euro 2000 Football Championships in Belgium and Holland. The Dutch set out to create a carnival atmosphere, removed objects that might be thrown from town squares and arranged for light beer to be served in plastic glasses. The Belgians threw tear gas grenades into crowded bars and arrested and deported large numbers of people who had done nothing wrong. And which country experienced problems with violence?

Consider also the pictures of a Premier League ground in Europe in Figure 5.4.1.

Safety and security at sports grounds 163

Figure 5.4.1 Security at the expense of safety and comfort.

Notice the lovely away end from which the seats have been removed so that the fans have to sit on the bare concrete. Notice also the delightful view from behind the fence. It was fences like this one which helped cause the Hillsborough disaster in 1989 and it's dreadful to still see them in Europe twelve years later.

Hillsborough led to radical change in the United Kingdom. There has been a paradigm shift from the management of public order to the management of public safety, within which disorder is just one of the risks to be dealt with. And we have seen the evolution of a complex framework of safety and security regulation. I tried to analyse it in about 1995 when it looked something like Figure 5.4.2 on the next page.

The point is to convey in a picture the huge number of organisations who have got involved in making recommendations about safety and security. Sports grounds have become one of the most regulated areas of public activity – a deluge of rules to replace past laxity and neglect.

FIFA, UEFA and the various national Football Leagues all have their own wide ranging and different requirements and rules for crowd management. Referees have to decide whether the match is on and then take responsibility for what happens on the field. The Football Licensing Authority have previously issued a licence for the ground to be allowed to admit any spectators at all – at Chesterfield, failures to carry out renovations resulted in the FLA closing part of the ground. Local authorities then issue a certificate to say how many spectators can be allowed in to the ground.

At least three different UK Government Departments are involved in regulation – Environment for Building Regulations, the Home Office for policing and fire safety and Culture, Media and Sport for safety guidance and legislation. The Association of Chief Police Officers make national policy on policing sports grounds. Local police decide on match categorisation (low, medium or high risk), 'all ticket' matches and kick-off times. And we have seen the emergence of a profession of ground safety officers, supported by their stewards. All of these organisations regularly attend events to carry out 'inspections' and meet together to discuss matters in the Safety Advisory Groups set up within each local authority area.

It was a Government Committee in 1991 which recommended that 'higher profile stewarding supported by lower profile policing' represented the way forward in sports grounds. The police had become concerned about their exposure to civil and even criminal liability in the event of another disaster – they did not want to take responsibility for safety just to fill the void left by nobody else accepting the responsibility. They did not want to be sued for negligence or face criminal charges for manslaughter. The police were faced with burgeoning demands for their services and public events offered one area where they could cut back on their involvement to free up resources to service new demands elsewhere. And although they could charge clubs for their services, often they could not recover all their costs.

Safety and security at sports grounds

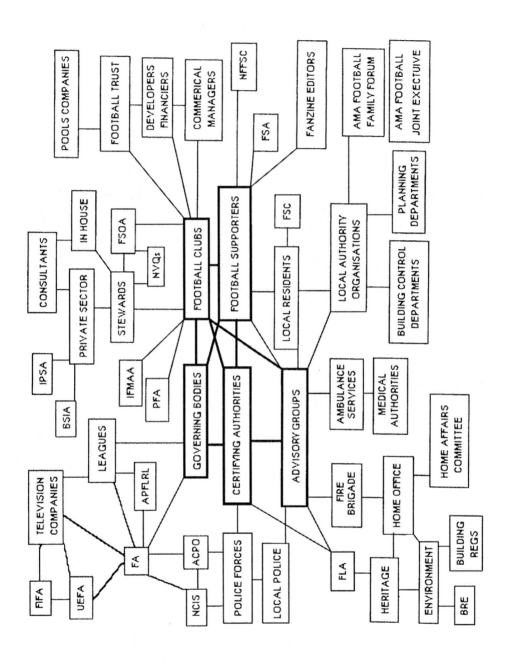

Figure 5.4.2 Complex framework of safety and security regulation.

So the police refocused their role. They appointed Intelligence Liaison Officers for each club, established a network of 'spotters' to monitor hooligans and set up a national football intelligence co-ordinating unit. They drew up statements of intent with clubs to clarify who was responsible for what – the club for safety and the police for order and emergencies. So clubs had primacy for 'policing' functions and called in the police to help when serious disorder or a major incident took place.

Following Government recommendations, ground management assumed full responsibility for the safety of their customers. Football led the way with the appointment of ground safety officers. In the early days, these were often the retired local police chief working part-time. Increasingly, however, we have seen the emergence of full-time, younger personnel with a health and safety rather than police background.

The Football Safety Officers Association was set up in 1992. Almost all safety officers are members. The FSOA has slowly grown in stature and influence and its national officers have worked with the various regulatory bodies on various developments such as stewards training and assessment. Other sports have also begun to appoint safety officers and sports such as rugby union and rugby league have set up their own Safety Officers Associations.

There is a growing debate about the competence of safety officers and there is a trend – painfully slow – towards greater professionalisation. Safety officers should assume full responsibility for all health and safety functions – staff and banqueting guests as well as the matchday crowd – and they need to demonstrate their competence through continuing professional development and qualifications.

So we have seen the police and stewards change places. I always quote the example of Nottingham Forest. In 1989 they used to have 150 police in the ground, supported by 75 stewards. For their UEFA Cup run in 1996 they had 250 stewards supported by just 22 police. And many clubs have no police officers at all in the ground. Portsmouth, for example, a [then] division one club – had only two matches last season at which there were police in the ground. However, there are usually police officers on duty outside the ground who the club can call on if needed. A good number of clubs now invite the visiting club to send their own stewards so that the 'away' fans are then policed by people who know them. Revised and updated guidelines for stewarding have been produced, together with a national training package and assessment scheme. The training package covers the general duties of stewards; their work to keep the stadium safe; caring for the spectators; giving medical first aid; preventing and dealing with fires; and emergency evacuation. It is used at almost every ground in the UK and has also been drawn on by other countries hosting major championships.

So having reviewed the changes, let's now reflect on their impact.

From the commercial perspective, there have been cost savings – stewards are substantially cheaper than paying for police officers. And there is better risk management. Taking proper responsibility for safety reduces exposure to the risks of liability. Safety officers and stewards can also get involved in a broader range of risk management activities than the police – for example supervising licensed bars or patrolling the stadium to look for fire risks.

On the downside, too many clubs look to cut costs beyond what is acceptable. One safety officer recently could not get his club to pay for the legally required annual testing of various technical installations. Another safety officer's club repeatedly sought to cut the number of stewards on duty below the level legally required. Yet football is awash with television money and many players earn obscene salaries.

Of course higher profile stewarding is confined to inside the stadium and its approaches. So any problems from noise, traffic, litter and vandalism experienced by stadium communities are largely unchanged. The evidence suggests that police and stewards, particularly with the support of Closed Circuit Television, have been very effective at reducing incidents of disorder inside grounds. But the evidence also suggests that the problem has been displaced away from the grounds, for example in town centres or on public transport systems.

Inside the ground, good stewarding is preferable to over-zealous policing. The emphasis of stewarding is on looking after people rather than treating them as the enemy. There are good examples of safety officers and stewards trying to facilitate fans enjoyment – allowing bands and banners under controlled conditions for example – when the police would simply say 'no'.

But there are issues about the accountability of stewards. For example, if you are wrongly ejected or arrested by a police officer there is an established system for making formal complaints. But if you get thrown out by a 'search and eject' team of stewards, seeking redress becomes rather more difficult.

There is also an argument that control couched in the language of safety is just as tyrannical as confrontation. The police used to say 'you will do this because I am telling you to', whereas stewards now say 'you must do this in the interests of your own and others' safety'. Same difference. Your behaviour in the stadium is still very strictly controlled.

I've already mentioned the vast reduction in hooliganism inside the ground. And it's also right that there has been only one safety/security related death inside a British ground since Hillsborough – at a Welsh international football match when a spectator was killed by a flare fired across the stadium. It is the focus on stewarding safety rather than policing disorder which has been a major factor here.

But lower profile policing means that police experience of sports events is reduced. Thus for matches where a police presence is needed, particularly the police commanders may be less competent and their decision making less capable.

There are also concerns about whether these welcome reductions in disaster and disorder are allowing complacency to creep back in. Standards of stewarding and safety management are still variable. The fact that Europe has not recently experienced a major disaster should not allow certain clubs to 'get away' with not putting their houses in order.

Countries which have staged major Championships have learned from the British experience. But in too many places, we still see high fences and draconian policing. We are still seeing injuries and deaths resulting from the indiscriminate and wholly inappropriate use of baton charges and tear gas. In too many countries, the policing of sports events is still about protection *from* the crowd rather than the protection *of* the crowd. The implementation of good safety management and stewarding systems could help change this for the better.

5.4 – COMMENTARY BY JIM CHALMERS – HIGHER PROFILE STEWARDING

The article written by Frosdick in 2001 evaluates the police withdrawal from football grounds in England and Wales and their replacement by improved stewarding and safety management. Frosdick questions whether this experience can be translated to other countries and contrasts UK policing styles with those employed elsewhere in Europe.

He resurrects his examination of the use of high pitch perimeter fences to segregate spectators from the playing area and uses pictures of a European stadium to make the point. However, whilst I agree such fences have no place in the modern stadium, I question whether we should be so critical of their use in Europe whilst Shrewsbury Town F.C. retains one such fence. I find it equally dreadful that, sixteen years after Hillsborough, one English club still has this relic from the past.

Frosdick writes accurately about the high level of police resources used to control football crowds in the 1980s. Going further back, I remember the 60s and 70s when policing football was more of a war with the fans than an enjoyable afternoon at the match. Despite all of the inquiries into football disasters and the legislation which clearly placed the safety responsibility at the door of the football clubs, their apathy and lack of effort resulted in the police being gradually sucked in, using extensive resources to fill the vacuum in crowd management. I speak from first hand experience as the Police Commander at Aston Villa. During each event, I saw the Chief Steward twice during a game, but the event was totally managed from the Police Control Room. There was little or no input from the Club on how the seated areas and terracing were managed. It was not until after the Hillsborough disaster that my 'Police Control Room' became the 'Stadium Control Room' and the Club Safety Officer became the event commander. Thus, from the 60s to the 80s, it was a case of 'high profile policing', sometimes supported by 'low profile stewarding'.

In this and other articles Frosdick refers to the growth of the Football Safety Officers' Association and the development of professional and competent safety officers and stewards. Earlier I referred to how the police had filled a vacuum at football grounds. To have just withdrawn the police would have left a huge void so at Aston Villa, one of the first objectives was to employ a safety officer. This role went to an ex senior police officer. He in turn employed younger, fitter and more capable stewards with a programme of training introduced long before that prepared by the football authorities. As their experience and confidence grew, I responded by withdrawing the police from inside the ground until the ideal level was reached when stewarding was at an all-time high and policing was at an all-time low.

This process of policing level adjustments has continued to the present day with approximately 60% of matches in the Premier and Football Leagues being 'police-free'. In the case of Kidderminster Harriers, we will have only five home fixtures policed during the 2004/05 season. Such a prospect would have been unthinkable in 1989, but subsequent changes have enabled the police and regulatory bodies to have confidence that clubs can successfully manage their stadia. This achievement is testimony to the dedication of the Safety Officers and their stewards, who are unsung heroes and heroines of every match.

At many high profile fixtures, a large number of police officers is still required to maintain public order and protect the community. This is evidenced by the Southampton and Portsmouth case study in Part IV, which is sadly not unique. However the likelihood of crowd trouble inside the stadium has been considerably reduced. Even when it does occur, it is normally left to the stewards to respond in the first instance, with the police attending in support if necessary. Contrast this with the situation prior to 1989, when the response to any crowd incident was left entirely to the police, with stewards very often nowhere to be seen.

Control mechanisms in the stadium have largely displaced crowd problems to outside and away from the ground. So whilst the ideal mix of policing and stewarding has been achieved inside the stadium, the police are still needed to respond to public order and public protection issues elsewhere.

I do think Frosdick has fallen into a trap of considering the changes in safety management and stewarding in isolation. Whilst all of them are significant, their impact on crowd management and behaviour could not have been achieved without the significant investment in stadium structures seen in the post Hillsborough era. When the new Wembley stadium is completed it is estimated that over £2 billion will have been spent on new and redeveloped stadia in England and Wales. Without the significant improvements in stadium design and facilities, it is unlikely that our attitudes to fences, policing, safety management and stewarding would have changed very much from the pre Hillsborough era.

I agree with Frosdick that, in many other countries, the former UK attitude still prevails of 'protection from the crowd', as distinct from 'protection of the crowd'. There is thus a need for the whole UK experience – structures and management – to be shared with other countries. It would be arrogant to suggest that what works in the UK would necessarily work in other countries. By all means let us be proud of our achievements and share our lessons and experiences, so that other countries may consider their applicability in the context of their own circumstances.

5.5 – DANGER: MEN AT WORK

The original citation for this article is: Frosdick, S. and Rankin, R. (2001) 'Danger: Men at Work', *Stadium and Arena Management*, February 2001, pp. 33-34.

Steve Frosdick and Bob Rankin reflect on the tragic death of a Security Guard/Steward at Coventry City Football Club

Rodney ('Ron') Reeves was killed whilst working at Coventry City's Highfield Road stadium on 31 October 1998. His death gave rise to various legal proceedings and, now that these have been concluded, we want to take the opportunity to reflect on events. There are important lessons to be learned and we hope that sharing them may help prevent other incidents elsewhere.

Previously a steward at Aston Villa, Ron came to work as a steward at Coventry City for the start of the 1996 season. He was the very epitome of customer service and thus well suited for his role as the Directors' Steward – 'meeting and greeting' Directors, VIPs and guests, including the visiting team, at the Main Stand entrance to the ground. In 1997, Coventry City set up its own full time security team. Ron was taken on as the supervisor of this team, but still retained his former match day role at the Directors' entrance.

The turnstiles are opened to admit the public 90 minutes before the start of any match and the away team coach should arrive at least one hour before the start of the event. There are usually supporters, in varying numbers depending on the popularity of the away team, gathered round hoping to catch a glimpse of this or that famous player. The coach, escorted by police motorcyclists, would be driven in from the main road, entering through some outer sliding gates onto club land fronting the Main Stand and turnstiles. It would then reverse through the Directors' gates into the Directors' car park to back into its parking space, where the players would disembark for their dressing room. So it went for every match for over twenty years – and nothing happened. And because of the considered adequate arrangements in place there was no perception of risk. This failure of hindsight presents our first salutary lesson. Risk assessment is inadequate if it is confined to putting things right only after they have gone wrong. It's easy to spot a risk once it has actually happened. But active foresight is equally important. So every aspect of the management system needs to be systematically considered. The hazards associated with each activity need to be identified and the risks assessed and managed.

On 31 October 1998, the visiting team was Arsenal – a very high profile club – and there were a number of supporters waiting to see their heroes. The coach was late, arriving at about 1355 hours. The sky was dark and it was raining heavily. Ron Reeves was wearing his dark coloured security guard's uniform and, because of the bad weather, this also included a black storm jacket and leggings. And here we find our second lesson – could Ron be seen? Stewards normally wear a high visibility reflective jacket, not only so they can be readily identified by the public, but also for their own safety. If they are dealing with an incident in the middle of the crowd, it is important that they can be monitored. And it's vital, if they are in an area of traffic movement, that they can be seen easily by vehicle drivers.

At the inquest, which recorded a verdict of accidental death, the Arsenal coach driver told how he had driven in from the road, swung round, checked his mirrors and begun to reverse. He had adjusted the steering so as to miss the gates and then seen a 'black smudge' in his mirrors. Suddenly people were banging on the coach. The driver stopped, pulled forward and saw a body shape drop to the ground. Ron Reeves had been seriously injured, crushed against the gatepost to the Directors' car park. Nobody knows how he came to be behind the coach. Many theories have been given, perhaps he had seen the gate start to swing, or perhaps he had been trying to squeeze through the gap. Most probably, and notwithstanding that he couldn't be seen, he was trying to help the coach driver avoid the gates.

Lesson three involves planning and supervision. All roles should have a written job description setting out duties and responsibilities and all staff should be regularly supervised. The role Ron was carrying out was a matter of custom and practice. It was not his job to assist the coach in – no steward or employee had ever been told to help the coach in, but nor had anybody been told not to. Nobody could say whether Ron regularly gave assistance or not. Unfortunately it is human nature to try and help a driver having difficulty steering – how many times have we all seen it?

Just after 1355, the accident was reported to the Control Room. With still an hour to kick-off, the medical resources required for the match had not yet confirmed they were in attendance, and the Control Room could not locate them. Lesson four – ensure that your Control Room is notified immediately when the agencies (medical, fire, etc.) arrive and where they are. When nothing ever goes wrong, it's so easy for the operation to become familiar, for complacency to develop and for sloppiness to creep in. Fortunately, there was an ambulance which had brought a terminally sick child to see their last game, and also the club doctor walked through the gates just as the accident happened. So medical help was in fact quickly available.

The last hour before kick-off is the Control Room's busiest time and the Safety Officer and his staff continued to supervise the safe admission of the spectators. Shortly after the match started, the Control Room was informed that Ron had died. The media picked up this up immediately and the fact that a steward had been killed was being broadcast nation-wide almost straightaway. The Safety Officer was immediately concerned for the families of the 200 stewards who worked at the ground and also realised that the switchboard could be deluged, therefore a priority media release was made that the deceased was a full-time member of the club's security staff. The importance of good record-keeping in respect of all staff was then brought into focus with difficulties tracing his next of kin – updated information had not been supplied to the club.

Providing support proved a considerable demand in the immediate aftermath of the accident. Ron's family were understandably devastated. The club kept them fully informed and provided help and support with the funeral and subsequent memorial service. In due course, collections were made at football grounds across the country, necessitating the setting up of a bank account. The two clubs, Coventry and Arsenal, also contributed substantial amounts to this fund for the family. The media were very hungry for information and the Safety Officer spent most of the first weekend dealing with their enquiries. Ron's stewarding and security colleagues were also badly affected. The Safety Officer, with the assistance of the local Council, prepared a leaflet on post traumatic stress disorder, giving staff contact telephone numbers and offering access to counselling services. About six members of staff and the public subsequently accepted such support.

The police investigated on behalf of the Coroner and took a report of the accident but, since it involved a member of staff, the accident also had to be reported to the relevant authorities in compliance with Health and Safety legislation. The Safety Officer therefore contacted the Environmental Health office of Coventry City Council via an emergency phone number. (This phone call was subsequently confirmed with a written report – known as a 'RIDDOR' report). A Health and Safety investigation was launched and Environmental Health officers interviewed the club Chairman, Company Secretary and Safety Officer. The interviews raised four particular issues. These were a perceived lack of segregation between vehicles and pedestrians, the fact that Ron had not been wearing a high visibility jacket, the narrow width of the gates and the documentation of Health and Safety procedures.

A further salutary lesson concerns the importance of documented risk assessments and acceptable Health and Safety procedures. These are not just good operational management practices, but also support accountability and so mitigate against liability. There is a varied approach to defined areas of Health and Safety responsibilities at football clubs right across the country. At Coventry City, the Safety Officer is responsible for spectator safety on match days and security issues. General health and safety at work was the responsibility of another staff member – the Stadium Manager – who was also considerably occupied with other duties.

Subsequently, a letter arrived from the Environmental Health office to point out the perceived deficiencies and to warn of a possible criminal prosecution. In due course, Coventry City FC were summonsed to appear at the Magistrates Court to answer five criminal charges:

- failing to ensure the Health and Safety at work of an employee;
- failing to ensure the Health and Safety of persons not in the club's employment;
- failing to organise a safe traffic management system;
- failing to provide suitable personal protective equipment; and
- failing to provide a risk assessment.

The club initially offered a plea of guilty to two of the charges, but this was not accepted by the prosecution. Instead, a committal to the Crown Court (a higher Court) was sought, where the judge would have unlimited powers of punishment.

Some months later, the case was listed for trial at the Crown Court. The club's lawyers tendered 'not guilty' pleas to the five indictments and then, pre-trial, submitted legal argument that the indictment was overloaded (five charges was excessive, imposed an undue burden on the jury and was therefore contrary to the interests of justice). The judge agreed and stayed the proceedings in respect of the three minor charges. The club's lawyers were then able to argue that the two remaining charges should also be dismissed, on the grounds of abuse of process. It had become apparent that there had been deficiencies in decision making within the local authority. The Council's Environmental Health office had not consulted with the Council department responsible for issuing the ground safety certificate and had also conducted a private hearing – at which the club had not been represented – before deciding to prosecute. After taking evidence from a senior local authority officer (responsible for issuing the club's safety certificate) and a lengthy legal discussion, the judge took the view that there had indeed been an abuse of process and there were compelling reasons why the criminal prosecution was flawed.

On the positive side, the case highlighted issues where improvements could be made in the area of Health and Safety. The club appointed consultants who undertook a total review. Procedures and personal record keeping were tightened up and risk assessments undertaken for all areas of work activity. Concerning the arrival of the away team coach, after the accident the club immediately set new 'sterile receipt' procedures in place and these received favourable comment in the legal proceedings which followed. Other clubs, quite rightly, have also taken steps to segregate vehicular and pedestrian traffic as a result of this accident.

The final lesson to be learned is that, whilst Coventry City has always considered the issue of safety paramount, and has always put in place any change recommended without flinching at the cost, any business needs professional and sound advice on matters of Health and Safety. At many clubs where resources are small, the Safety Officer takes on the mantle of 'Health and Safety' purely because of the word 'Safety' in their job title. However, they may not have the necessary background or qualifications for the role. Please be careful. In cases such as this one, English law switches the burden of proof from the prosecution to the defence.

5.5 – COMMENTARY BY JIM CHALMERS – ACCIDENTAL DEATH OF A STEWARD

The Article written by Frosdick and Rankin in 2001 followed the tragic death of Ron Reeves prior to the Coventry City fixture against Arsenal at Highfield Road, Coventry, in 1998. I knew Ron well from my involvement with his club as an Inspector with the Football Licensing Authority. The memories of his death remain with me today, as they do with all of us privileged to know him.

The article is testimony to the honesty and openness of Bob Rankin, the Safety and Security Manager at Coventry City FC at the time of Ron's death. I particularly remember Bob's courage in sharing these experiences at an annual general meeting of the Football Safety Officers' Association, when he openly admitted the mistakes that were made with health and safety procedures and warned his colleagues on how to avoid them.

Since 1991, as an FLA Inspector, I had monitored the discharge by Coventry City Council of their statutory obligations under the safety at sports grounds legislation and was still in this position in 1998. I never had any doubt as to the commitment, professionalism and integrity of the council officers directly responsible for discharging these responsibilities in respect of Coventry City FC. At numerous safety advisory group meetings, this was a view shared by the emergency services, the voluntary aid agency and the supporters' representative. I was similarly impressed by the commitment of the club chairman, directors, stadium manager and Bob Rankin to the care and safety of everyone attending an event at their stadium. Over the years, the club was able to prove by example that the stadium structures were built, maintained and managed to meet the relevant spectator safety legislation and guidance.

However neither I nor the council officers were involved with or had any statutory duty for monitoring or enforcing the employer/employee health and safety legislation. Given the impeccable record of the club in respect of the safety at sports grounds legislation, there was no reason to doubt a similar commitment to employee health and safety. That is why I was always surprised that the Environmental Health Department of Coventry City Council never interviewed me as part of their investigation since I always felt that my evidence in respect of monitoring safety at the stadium would have been relevant in deciding whether a prosecution was warranted. The potential of this evidence was not lost on the Club defence solicitors who if the case had gone to trial would have called me as a defence witness.

Despite the club's excellent safety record, the tragic death still occurred during the parking of the visiting team coach, a scene that had been repeated over many years without incident. Such a scenario was being played out up and down the country at various clubs with each club no doubt thinking that their employee health and safety responsibilities were being complied with. Certainly Coventry City did not perceive that the coach manoeuvre was a health and safety risk. This demonstrates that even in the best run organisations, no one can ever be complacent about discharging their legal liabilities for the health and safety of their employees.

The tragic death provided a wake up call to all clubs, including those in my region, when there was a flurry of activity to put in place a traffic management plan, to review the personal protective equipment issued to car park personnel, to conduct risk assessments of car parking duties, to train personnel in such duties and finally prior to each event to brief them as to their responsibilities, including ensuring their own health and safety during the movement of vehicles.

I think it true to say that most clubs would have fallen foul of the breaches of health and safety legislation identified following Ron's death, but Bob Rankin's sharing of his experiences helped clubs put their house in order before it was too late. Nevertheless, I know of clubs which, even today, ignore their health and safety responsibilities for car park personnel – despite the consequences for them and their staff if things go wrong.

The article still provides a useful case study of what can go wrong within even the best-ordered organisation. I know the chairman, directors and staff at Coventry City FC will never forget the trauma they went through after the death of such a well-liked and popular employee. The article should be made compulsory reading for every sports club chairman, director, chief executive and safety officer. Ron's death should serve as constant reminder of not only the tragic personal consequences, but also the legal consequences, both criminal and civil, if clubs do not act with due diligence in their health and safety at work responsibilities. The case reinforces the fact that complacency is the main threat to safety.

PART VI – SEATING

Introduction

Part VI comprises five articles relating to the introduction of all-seated stadia in the aftermath of the 1989 Hillsborough stadium disaster. The first two articles examine the lack of atmosphere in an all-seated stadium. The next two articles consider the ongoing debate about the reintroduction of standing terraces and look at the problems caused by fans who persistently stand in seated areas of the ground. Part VI concludes with a reflection on the tenth anniversary of all-seated stadia.

6.1 and 6.2 – Atmosphere in All-Seated Stadia

This pair of articles were written in 1997 and 1998. They examine the effects on crowd atmosphere of an all-seated stadium environment. Article 6.1 discusses the impact at Ewood Park (home of Blackburn Rovers FC) and gives a case study of the efforts made to improve the match day atmosphere at that football ground. Article 6.2 is more of an academic explanation of the factors which can influence stadium atmosphere and gives examples from other sports and countries. Comment is made that atmosphere is only one aspect of the event experience and should not be viewed in isolation from more important issues such as safety, comfort and enjoyment of the game. Comment is also made that present and future generations of football fans may not necessarily have the same opinion of match day atmosphere as the traditional supporter.

6.3 and 6.4 – Persistent Standing in Seated Areas

Both articles written by Frosdick in 1998 and 2004 examine whether there is a case to be made for the re-introduction of standing terraces in the top two football divisions in England and Wales. Article 6.4 specifically discusses the phenomenon of fans who persistently stand in seated areas of football grounds as a direct consequence of the introduction of the all-seated stadium. Both articles refer to the ongoing debate on whether safe standing terraces can be re-introduced, with arguments for and against this proposal. Comment is made that a distinction needs to be drawn between the argument for the re-introduction of standing terraces and the safety issues which stem from fans who persistently stand. Comment is also made that it is unlikely that standing terraces will ever be re-introduced in the top two football divisions, thus the problem of persistent standing will remain to be dealt with. The scale of the problem and measures introduced to combat it are also commented upon.

6.5 – Ten Years of All-Seating

Writing in 2004, Frosdick gives a brief review of ten years of all-seated stands in England and Wales. Comment is made on the practical issues faced by Clubs and the Football Licensing Authority during those ten years. These include the enforcement of the all-seated requirement and the achievement of high standards in stadium development. Comment is made that the stadium changes went beyond the provision of seating to include improved quality of viewing, facilities and amenities. Reference is also made to the many new stadia which have been built and to how this redevelopment process will continue in the future.

6.1 – EWOOD EFFECT

The original citation for this article is: Frosdick, S. and Highmore, M. (1997), 'Ewood Effect', *Football Decision*, Issue 7, August 1997, pp. 20-24.

The atmosphere at football grounds is not what it was. Steve Frosdick and Mel Highmore report on the problems, the recommendations of an FA Premier League Working Party and the action being taken at Blackburn Rovers' Ewood Park stadium.

Since the implementation of the Taylor Report, all Premier League and First Division clubs in the UK have been required make their stadia all-seated. An immediately apparent effect of this was that social groupings on the terraces were split apart. Where once the fans had joined arms, swayed, sung, chanted, celebrated and commiserated with each other, suddenly they were no longer able to do this. The effect was particularly marked in stands where you had to sit in a designated seat rather than choose where you wanted to sit or stand. Yes, the seats were comfortable, yes, less injuries were sustained and yes, you were assured of your favoured viewing position if you had a season ticket. But something seemed to be missing.

Concerns about crowd atmosphere first 'officially' appeared during discussions held at FA Premier League Supporter Panels during the 1995/1996 season. Panel members were saying that the atmosphere was less exciting than it used to be; that a good atmosphere was very important to both the fans and the players; and that something needed to be done, club by club, to make the atmosphere better.

Respondents to the 1996/97 Premier League Fans Survey carried out by Leicester University made it clear that they were dissatisfied with the poor atmosphere at their stadia. 60% said the atmosphere was not what it was. 40% thought it was too quiet or dull and 20% definitely disliked their club's pre-match and half-time entertainment. Whilst opinions varied from club to club, the survey concluded that at least half the clubs needed to make considerable improvements.

The Premier League decided that this disaffection needed to be addressed, primarily because of their wish to respond to customer concerns, but no doubt also because atmosphere is a key feature in making football attractive to television. Cable and satellite viewers will know how dull the televised game when broadcast from a virtually empty stadium. BSkyB pay millions for the Premier League product, and they expect a bit of noise and colour along with the football.

Accordingly, the Premier League asked Paul Johns to co-ordinate a working party to investigate the problem and make recommendations. The working party visited matches, talked to clubs and supporters, and discovered further evidence that it was the social interaction of terraced environs which had been torn apart. In the interests of comparison, the working party visited Serie A in Italy and also took a brief look at rugby league and ice hockey. They worked from the premise that supporters' dissatisfaction with entertainment and atmosphere was important. Fans go to a match not just to watch but to participate. They want to be part of an event which begins on the way to a game, includes time at the stadium before the game, takes place in the stand and concourses as well as on the pitch, and involves the supporters as well as the players. Thus atmosphere is part of what they pay for.

The working party reported in March 1997 with the following general conclusions:

- the bigger the crowd, the better the atmosphere, thus regular supporters should be encouraged;

- 'atmosphere areas' – areas of the crowd naturally disposed to be noisy and colourful – should be built up in either unreserved or season ticket sections;

- away fans contribute to the atmosphere out of all proportion to their numbers, thus away travel should be encouraged;

- what the fans do when the players are on the pitch is at least as important as what the club does when they are not;

- before the game, club entertainment should be directed less at the inside of the stadium and more at the concourse and immediate environs;

- improving atmosphere and entertainment has implications for stewarding and may involve changing the rules about what fans can and cannot do;

- club initiatives are more likely to succeed if the club makes a point of consulting and involving fans; and

- clubs should not expect new initiatives to lead to instant improvement.

Mel Highmore, the stadium manager at Ewood Park, had been a member of the working party. His background in the police service and the emphasis of his health and safety work had previously made him inclined to refuse fans requests for atmosphere initiatives on grounds of the risks to safety and security. Instruments or small flag poles could be used as weapons, whilst flags and coloured papers posed a fire risk. Mel now recognised the need for change. Whilst actions to improve atmosphere clearly had take account of the requirements for crowd safety, the application of structured risk assessment methods would allow both for unacceptable ideas to be screened out and for safe and sensible initiatives to be introduced in a controlled way.

There were three good reasons for Blackburn Rovers to tackle the issue of poor atmosphere at Ewood Park. First, the club were still in danger of relegation and it was felt that a noisy dynamic crowd supporting the team would give encouragement to the players. Second, it was clear that, whatever the results on the pitch, a chanting, singing crowd would add increased value and enjoyment for the fans. Third, from a safety and security perspective, it was thought that the crowd would be better humoured and less aggressive if music and singing were encouraged. There were still three home matches left – so what was to be done?

Sheffield Wednesday have a band which always play at Hillsborough and which like to play away where this is allowed. Like many other clubs, Mel Highmore had always refused Sheffield fans when they asked to bring their musical instruments. But now he agreed to meet the band leader to discuss allowing instruments into the stadium subject to certain controls and consultations.

But clearly Rovers had to have a band too. Mel got to hear of Goodshaw Village Brass Band and managed to recruit four musicians – three trumpets and a trombone – to bolster the efforts of an unidentified drummer who all season long had conducted a solo performance from his seat in the Blackburn End Stand. Mel sought the opinion of several groups of fans. They certainly remembered the marvellous musical efforts of the Dutch supporters during Euro '96 and fully supported the idea of Rovers starting their own band. It was also important to obtain the agreement of Blackburn Rovers Caretaker Manager, Tony Parkes. After some discussion, he was persuaded that the band would bolster spectator support for the players.

It was important to ask supporters where the bands should be sited, as well as taking into account the safety implications – the musical instruments could not obstruct the gangways or exit routes. Thus it seemed that the band should ideally be located at the rear of a stand, in the 'engine room' of vocal support – behind the goal – with the stand itself acting as a giant megaphone.

Come the day of the match, both sets of musicians turned up as arranged and were escorted to the places which had been set aside for them: Sheffield at the rear of the away zone in the lower Darwen End; and Rovers with the drummer in the centre rear of the lower Blackburn End. To avoid alienating the fans, we wanted to avoid indiscriminate drumming and fanfares of brass, as well as those tunes which had indecent, racist or otherwise offensive lyrics. We were able to draw on the experience of the drummer – a long-time Rovers supporter. He had already discovered which tunes were favoured and understood that singing and chanting were spontaneous and linked to events on the pitch. Recognising that it was the fans who generate the atmosphere, not the club, the musicians were advised against trying and instigate the singing. Rather they should judge when supporters were starting to be vocal, then come in alongside them to boost their efforts.

The Sheffield band were the more experienced but the Rovers musicians soon got into the swing of it. There was music coming from each end of the stadium, and the fans were very responsive to it. As Wednesday attacked, their fans urged the team forward and the band joined in to support them. Then play would switch and now the Rovers fans and band took up the initiative. The pattern of play on the pitch was being reflected by the fans and the musicians. As a general rule at football matches, the fans sing loudly for the first fifteen minutes or so, then the atmosphere tails off until a goal is scored. This didn't happen at this match. Not only were Rovers four nil up before half-time, but the crowd noise was sustained throughout the whole of the match. And although there were only just over 20,000 people in the stadium, a relatively low crowd for Ewood Park, the singing and chanting was probably the loudest of the season. Even the normally more placid fans in the side stands were beginning to join in.

There was a real atmosphere of good humoured excitement about the place, and people were commenting to each other about it. The result apart, the undoubted highlight of the evening came when the Sheffield supporters and their band – four one down with fifteen minutes to go – began the chant the Dutch had used at Euro '96. The Rovers fans and their band took up the melody and suddenly the whole stadium was singing along together. Both sets of fans kept it up right to the end of the match and it was brilliant.

Following the match, the feedback received from both sets of supporters was excellent. Callers to 'Radio Rovers' were saying it was the best the atmosphere had been all season. Spectators were hugely in favour of the band being invited to play again. Even the club directors were enthused and suggested the purchase of a 'louder drum', which was duly obtained for the Middlesbrough match. The BBC TV camera position at the Blackburn End meant the band had to be relocated to a new site on the upper tier. Nevertheless, with the new drum in use, their presence again proved both popular and effective.

The final game was planned as an end of season party. Leicester City were more than pleased to bring their mascot – Filbert, their cheer leaders and a large lifelike model of their manager. These all toured the perimeter with the Rovers mascot and the cheerleaders formed a guard of honour for the players entry onto the pitch. Five hundred children were treated to face painting and every child was given a blue or white balloon to inflate. Flags and bunting were used to dress the stadium and spectators were encouraged to wear their 'home' shirts and to bring flags and banners. The Rovers band, once again, were active and successful in engaging the entire stadium in a succession of songs and chants. Whilst all these atmosphere initiatives may not possible or practical for every match, the final game showed once again that fans enjoying themselves and singing along together create a far safer and happy environment than one dominated by the bad tempered, aggressive and disorderly element.

Looking ahead to next season, Rovers want to build on this early success. They realise that atmosphere cannot be generated or imposed by the club management. This will be seen as a cynical attempt to cover other inadequacies. They appreciate that fans may need encouragement to become involved in atmosphere initiatives. This means building a degree of trust between club management and groups of fans. Only when the relationship is sufficiently developed will the fans co-operate and maybe agree to include their friends in the activity. Encouragingly, Rovers are already hearing from other club supporters who want to get involved with atmosphere activities when they visit Ewood Park next season. Let's hope that these efforts can act as a catalyst for improved atmosphere throughout the Premier and Football Leagues.

Figure 6.1.1 Different sports can learn from each other when it comes to crowd participation. Here the fans from Lithuania are as much an attraction as the game on the basketball court at the Athens Olympic Games.

6.2 – BEING THERE!

The original citation for this article is: Frosdick, S. (1998) 'Being There!', *Stadium and Arena Management*, Volume 2 Number 1, February 1998, pp. 6-10.

A good atmosphere is integral to a successful event. Steve Frosdick looks at the factors which can combine to provide patrons with an enjoyable and memorable sense of occasion.

Atmospheric Conditioning

Television brings an array of sports from across the globe into our living rooms. But for many sports fans, TV events lack one magic ingredient – the special atmosphere you only get from 'being there'. This sense of anticipation of occasion, of place and of being a participant in the event itself, is part of what patrons pay for when they choose to attend a live event. Equally importantly, since atmosphere is a key factor in making sports attractive to television, so broadcasters need to be provided with colour and excitement along with the event. Strategies to maximise revenue streams through gate receipts and through broadcasting rights need to respond to these demands. Good atmosphere is good business.

So what is atmosphere and what makes the atmosphere at a sports event? Psychologist Professor David Canter has described how stadium surroundings convey meaning to people. He says 'the experience of the facilities, the types of crowd behaviour that will occur and the management of a crowd will all take on different emphases in different locations'. So I want to suggest that atmosphere is a personal experience generated through the interaction of three environments; the architectural form, the crowd and the management. Let's look at these three areas and see what factors contribute to the atmosphere.

The individual architecture of each stadium or arena certainly makes an important contribution to the atmosphere. Writing of English football grounds in 1981, the anthropologist Desmond Morris told how 'the oddity of the shape of differing grandstands gives a sense of special location and provides each ground with its own characteristics'. Stadia are all architecturally different and sports fans love to travel to different grounds precisely to experience that uniqueness and diversity. Notwithstanding its dated facilities, the twin towers of Wembley stadium – the Venue of Legends – still act as a magnet to soccer fans and players across the world. Pele once said of Wembley, 'It is the church of football. It is the capital of football and it is the heart of football'.

Stadium writer Simon Inglis has pointed out that, for real sports fans, the 'ground – its stands, its terracing, its floodlights, its crowd, its noise, even its smell – is as much a part of the event as the match itself'. He speaks of 'stadiumitis' and tells how 'the merest glimpse of a stand, a floodlight or a pitch from a passing train or car, is often enough to set the pulse racing'. More and more stadia are enjoying additional revenue by providing sufferers with guided tours of the ground. Similarly, many operators partially floodlight the stadium to provide an impressive backdrop for banqueting events being held on non-match days.

The geographer John Bale, in 'Sport, Space and the City', shows how sports fans use metaphors to describe their special sense of place. Stadia are sacred places, referred to as 'cathedrals of sport'. They are 'home' to their fans. They are scenic spaces, named as 'parks'. And they are sources of heritage. On the US NBA website, one Celtics basketball fan describes his first visit to Boston Garden; 'It held so much history and character in its old walls, court and seats. The site of many championships and Hall of Fame players'.

At a Cost

There is a simple message here. High quality and distinctive design may cost more in the short term, but it contributes to a good atmosphere in the long term and a good atmosphere contributes to profits. Italy certainly seems to have understood this with what Owen Luder, Past President of the Royal Institute of British Architects described as the 'stirring contexts and breathtaking design' of the stadia built for the 1990 World Cup. These included Bari's elliptical San Nicola stadium, which from a distance looks like a huge flying saucer, and the Meazza Stadium in Milan, with its spectacular ramps spiralling round the perimeter walls. In 1992 British Channel 4 television screened a documentary contrasting Italian stadia with much of the 'container architecture' in English football. Here a cost rather than quality imperative led to redevelopments such as Walsall and Scunthorpe being described as drab soulless edge-of-town industrial units, with an atmosphere to match. But there is some good news in England. Football can now boast Arsenal's North Stand and the award-winning Huddersfield, whilst stylish developments elsewhere include Wimbledon Tennis and rugby grounds at Twickenham and Murrayfield. Derby County's new Pride Park provides an excellent example of a stadium designed for atmosphere. The crowd are near the pitch, but most importantly, the roof line has been designed to retain all the noise within the ground.

Approaching the millennium, a third generation of world class stadia, with retractable roofs such as in Toronto or Amsterdam, and moveable pitches such as in Arnhem, are coming along to provide a new awesome sense of place through leading edge technology and quite staggering feats of engineering.

The Crowd

So, the stadium invokes a special sense of place, even when empty. But the crowd adds considerably to the atmosphere. Different types of crowd behave differently, demanding different mixes of resources to provide for their safety and security. And different crowds have different impacts on the atmosphere. So there are good reasons for stadium and arena managers to try to understand their crowds. Management provided entertainments need to facilitate and complement what the crowd are doing. The wrong choices could do damage rather than enhance the fans experience.

Researchers have found it very difficult to classify the crowd. Some have tried looking at violent or peaceful crowds, or at dissenting or consenting crowds. Ibrahim Cerrah from the University of Leicester has suggested a fourfold classification of crowds as organised/unorganised and active/passive.

Crowds in the stadium may certainly be regarded as organised. They are to some degree constrained by where they can sit or stand; the event has a timetable and the security forces will seek to curb any excesses in their behaviour. Organised crowds are the stadium manager's staple diet across a range of events, including sports fixtures, pop concerts and political and spiritual rallies. What is key is the extent to which they are active or passive. Organised and active crowds create the atmosphere, whilst the organised and passive 'consume' it. Witnessing the noise and spectacle of the active participants is part of what the inhabitants of the sky boxes and club seats pay for. It's better than television, but they are still watching rather than joining in.

The provision of entertainments such as cheerleaders, bands and loud music will appeal to the passive, yet annoy the active. For example, in the Olympic Stadium in Munich, the public address system plays the can-can at deafening volume every time Bayern Munich score a goal. This infuriatingly stifles the fans' own celebrations. A recent English Football Association Working Party looking at the loss of atmosphere in all-seated grounds concluded that 'what the fans do when the players are on the pitch is at least as important as what the club does when they are not. It may be easier for clubs to provide controlled entertainment before kick-off and at half-time. But it may well be more productive to invest time and resources in encouraging fans to create their own atmosphere during the game'.

The Italians, again, seem to have learned this lesson by facilitating supporters' organisations such as the ultras to back their team with orchestrated noise, colour and movement. Great stadia and tremendous fans – little wonder Serie A Italian football has its reputation for atmosphere.

Crowds that are unorganised and active may be found in spontaneous protests such as stadium managers can encounter when the fans are angry with the club. For example, at the English football club Brighton and Hove Albion in 1996, there were frequent pitch invasions and pre- and post-match demonstrations protesting against the directors selling off the ground. Here the atmosphere was angry and hostile, and sensitive policing was needed to avoid provoking serious disorder.

Having recently visited the Oktoberfest in Munich, I take the view that beer festivals also fall into this category. An arena filled with uninhibited drunks, standing on tables above a floor awash with broken glass, is anarchic rather than organised. The superficial atmosphere may be jovial, but you can cut it with a knife. There is a sense of tension and outbreaks of violence are common.

Unorganised and passive describes the street crowd, including the majority of fans on their way to the stadium. They are mostly still consumers rather than participants at this stage, and their sense of anticipation will increase as they near the stadium. Thus venue managers should reflect on what they can do on the approaches and concourses to heighten the fans experience.

Management

Thus for both the active and passive sports fans, it is the pre-match build up and not the event itself which provides the venue with the best opportunity to help create the atmosphere.

On the approaches and concourses, it is possible to use a mixture of live action, images and audio. The FA Working Party suggested street theatre, bands, and – borrowing from Italy – the distribution of leaflets inviting people to join in whatever choreography the fans have planned to welcome the teams onto the pitch. Images can be shown on concourse television screens, Jumbotrons and even, as is planned for the Gelredome in Arnhem, in the buses conveying fans from the park-and-ride. Audio can include not only public address, but also a club radio station broadcasting over a radius of several miles.

By the Numbers

During the build up, it is good practice to use predictable signals and events which fans can anticipate. These help create the sense of the particular experience of watching that club. Some clubs announce the team in a particular way. The Bayern Munich host always says only the player's first name, and the crowd scream out the surname in reply. `Nummer Vier der Matthias … SAMMER'. As a German speaker, I was asked to host the Bayern fans for their UEFA Cup match at Nottingham Forest. We announced the team in the way they were used to, and got excellent feedback in the immediate roars of approval from the fans. Many clubs play a signature tune as the players take the field. Most spectacularly, at the United Centre in Chicago, the Bulls basketball team make their entrance to a hi-tech laser show.

During the event, as the experience of Italian football shows, facilitation and not imposition is the key to successful atmosphere for the more active crowds. Drummers can use rhythm to get the fans going, then bands can take up the refrain to support the fans' singing. In all-seated venues, special 'atmosphere areas' can be provided for those fans who want to be more active. And yet the American experience seems different. American football crowds are used to and appreciate the involvement of cheerleaders, bands, match commentators and even the match officials.

In ice hockey, there is a strong tradition of organ music during pauses in the play. My own experience is insufficient to say whether these crowds are less actively involved in the event itself, and thus more receptive to entertainments, but my guess is that Latin American and Southern European crowds are the more active and passionate, whilst the US and North European are generally more passive.

And finally...

Finally, for some fans, the attendance at the event may mark a special personal occasion, which in itself adds to their experience. It may be a birthday, or the first time father and son have gone together to support a team. The NBA home page asked basketball fans what they remembered about the first game they ever saw. The answers showed how the experience of the first match lives in the memory. Venues can both enhance these personal experiences and earn additional revenue through merchandising and involvement opportunities such as customisable souvenirs and even leading the team out.

Figure 6.2.1 The Japanese Formula One Grand Prix at Suzuka is a special event for these supporters of Japanese driver Takuma Sato.

6.1 AND 6.2 – COMMENTARY BY JIM CHALMERS – IMPROVING THE ATMOSPHERE IN ALL-SEATED STADIA

Article 6.1, written by Frosdick and Highmore in 1997, examines the issue of crowd atmosphere in the all-seated Ewood Park stadium, the home of Blackburn Rovers FC. Article 6.2, written by Frosdick in 1998, continues the same debate, exploring in more detail what is meant by 'atmosphere' and discussing practical experiences in other sports and in other countries.

The evidence in all of the articles in Part VI would tend to show that concerns about stadium atmosphere are inexorably linked with the introduction of all-seated stadia. I think Frosdick and Highmore should have presented a more balanced discussion on the benefits and disbenefits of all-seating. They have focussed on one narrow disbenefit. Other issues such as safety, security, comfort, enjoyment, behaviour and value for money are all linked to the stadium environment. These are equally if not more important than atmosphere, which is only one element in the overall event experience.

In Article 6.2, Frosdick explains the factors that can contribute to 'atmosphere'. I suggest this remains a very nebulous concept. Those of us who stood on the terracing from the 1960s to early 1990s will all have memories of the experience. Sometimes the atmosphere was one of sheer exhilaration and excitement. We were among thousands of other fans on a terrace with everyone acting in unison and singing in one voice. In other instances the atmosphere was one of sheer terror at being lifted off your feet in a surge down the terracing. Some occasions were jubilant whilst others were frightening due to the sheer malevolence of opposing fans that were just the other side of a high mesh fence. I will not go into the primitive toilet and refreshment facilities we had to endure with toilets being no more than an open sewer with urine running down the terracing and caged boxes dispensing cold pies and lukewarm Bovril. On some terracing with no roofs, in the winter you were soaked to the skin, although the crowd kept you warm, and in the summer you lost a few pounds in the terrace sauna. If all of this stirs nostalgic memories for the reader then this will certainly age you. Others who have never stood on a terrace will probably be asking what the fuss is all about. As I have got older I ask myself the same question.

In Article 6.1 Frosdick mentions the 1996/97 season Premier League fan survey carried out by the University of Leicester. In a similar survey carried out in 2001, fans were asked to consider various 'aspects of match days which have got better due to all-seated stadia?' With the exception of atmosphere, the responses were all positive about the benefits of all-seated stadia. 85% said fan safety had got better, 77% said comfort levels were better, 72% said the view of the match was better, 69% said supporter behaviour had got better but only 13% said the match atmosphere had got better. There were, however, wide variations between clubs. At one club,

82% of fans surveyed thought the atmosphere had worsened with the introduction of all-seated stadia. At others only four out of every ten fans surveyed thought that seating had made the atmosphere worse. Considering the other benefits perceived by fans, perhaps a changed stadium atmosphere is a small price to pay.

The contemporary football audience is far different to that of yesterday. It is said that for the 1966 World Cup Final at Wembley Stadium only 4% of the crowd was female. Yet for the 1996 European Championship Final, also at Wembley, the female proportion was estimated at between 35% and 40%. Men no longer dominate the football stadium and this changing audience profile has itself contributed to the different atmosphere in the ground.

I think that Article 6.2 provides some of the answers as to how the football fan can generate a positive, if different, atmosphere. I do not think the atmosphere of a large standing terrace can ever be replicated – but nor do I think that many people actually want this. Take for example the home FA Cup fixture in the 2003/04 season between my club, Kidderminster Harriers FC, and Wolverhampton Wanderers FC. Molineux Stadium, the home of Wolverhampton Wanderers FC, has been an all-seated stadium for many years and this game presented many with their first chance to stand on a terrace. Over 2,000 visiting fans snapped up the terrace tickets before the seats had sold out and I spoke to many during and after the game. Many of the older fans said the experience had taken them back and was a trip down memory lane. I asked some of the younger fans if they would swap their modern all-seated stadium for our modest standing terrace. The answer was a resounding 'no'. Interestingly, the visiting fans in the seated area made just as much noise as those on the terrace so there was nothing to choose between the 'atmosphere' in either.

There is still passion – excitement, noise, music, chants and songs – in the all-seated stand. No one will pretend that the unity of purpose generated by a large standing terrace is present. However, I suggest that fans welcome the improvements to the stadium. They welcome the enhanced safety and security they now enjoy. They welcome the improved facilities and amenities. They welcome the improved view they have of the match. So I suggest that the issue of 'atmosphere' is being used as part of the campaign for the re-introduction of standing terraces, which will be debated in articles 6.3 and 6.4.

6.3 – STANDING UP AGAIN?

The original citation for this article is: Frosdick, S. (1998) 'Standing Up Again?', *Football Management*, March 1998, pp. 8-10.

The recent problems at Old Trafford have reopened the debate about standing terraces at football grounds. Steve Frosdick assesses the security and safety implications.

Some months ago, Martin Edwards of Manchester United mooted the idea of reintroducing terracing at all-seater football grounds. While Edwards has since withdrawn his remarks, current events at his own club have again drawn attention to the question.

Andy Walsh, from the Independent Manchester United Supporters' Association (IMUSA), explains the problems at Old Trafford. 'About three years ago at an Arsenal match, supporters in the East stand stood up and began singing to 'get behind the team. A tannoy announcement asked people to sit down. After the game, people felt the gentrification of Old Trafford had gone too far. This was the catalyst for the formation of IMUSA. We have campaigned to be allowed to stand and sing to improve the atmosphere. We appreciate that some fans don't want other people jumping up and down in front of them. We know there's been friction, so we've persistently asked for a small defined singing area, so far without success.'

Most clubs had initial problems persuading some fans to sit, but good stewarding eventually succeeded. United, however, have tolerated supporters who continued to stand. Eventually, as United secretary Ken Merrett explained to the press, 'Trafford Council have told us that we risk having our capacity reduced if supporters do not remain seated for the main part of the game. We have advised supporters and we've taken a very relaxed attitude as far as we can, but drastic measures are now called for.' Fans have complained about heavy-handed stewarding and on Boxing Day, in the match against Everton, there was fighting between fans and the security staff brought in to eject them from the ground.

All this conflict could be easily resolved if the club were to reintroduce a small standing terrace. So what is stopping them?

There have been at least 41 crowd disasters or incidents in British football grounds since 1896. Crowd pressure, either direct or leading to structural collapse, was the immediate cause of all but two of them (one fire and one disorder incident). Crowd pressure happens on standing terraces and this is why Lord Justice Taylor became convinced that their elimination would be the single most important measure for improving safety.

The evidence suggests that Taylor was right. John de Quidt is Chief Executive of the Football Licensing Authority. He says that 'Seating is inherently safer than standing. I always remind people that, before the Kop at Liverpool was rebuilt, some 25 to 40 people were injured every match. Now only three to five are treated throughout the whole ground, mostly for illness rather than injury.'

Graham West, stadium safety officer at Loftus Road, adds that, 'Anyone considering a return to terracing has clearly forgotten the lessons learned. The safety benefits of all-seater far outweigh standing, particularly since there is no crowd surging in response to incidents on the pitch.'

Blackburn Rovers safety officer, John Newsham, is a firm fan of all-seating, but points out that it is not without its own problems. 'Supporters have lost a lot of choice. When buying a seat or a season ticket, you don't quite know where you will be or who you will be next to. On terraces, you could move to get a better view or escape from a loudmouth. In reserved seats you're stuck. It's also much harder for police and stewards to get into a seated area to eject or arrest an offender.'

There are also safety problems related to crowd dynamics, as West explains. 'A seated crowd stands up in response to excitement, and then sits down again. But if the crowd is already standing up, crowd dynamics mean they will surge forward. On terraces, they could sway forward then back again. But I am aware of at least one incident where a crowd standing up in a seated area surged forward when a goal was scored, fell over the seats in front and suffered injuries.'

Furthermore, all-seating is now recognised as the main factor behind loss of atmosphere at matches. Social groupings on the terraces have been split apart, particularly by reserved seating. Fans who once joined arms, swayed, sang, chanted, celebrated and commiserated with the same people week after week, have found it impossible to recreate this atmosphere with the strangers around them.

But these new problems are far more easily resolved than an excess in crowd pressures. Police and stewards have learned not to wade into seated areas, which almost always turns a minor incident into something worse. Closed-circuit television (CCTV) and reserved seating makes the identification of offenders much easier, allowing action to be taken at a more opportune time. Many an offending fan has failed to return from a half-time visit to the toilet, or even found the police calling at their home next day. Stadium managers are also learning to deal more sensitively with complaints from fans, for example, using covert stewards to investigate allegations of foul language. Stewards at most grounds have won the battle of getting the fans to sit. And the wave of reconstructions means there are far fewer examples of poor viewing quality caused by simply bolting seats onto an old terrace.

Even the atmosphere problem is beginning to be addressed. Last year, an FA Premier League Working Party concluded that 'what the fans do when the players are on the pitch is at least as important as what the club does when they are not. It may be easier for clubs to provide controlled entertainment before kick-off and at half time, but it may well be more productive to invest time and resources in encouraging fans to create their own atmosphere during the game'.

As the Football Safety Officers' Association chairman, Leon Blackburn, concludes, 'There is lots of evidence that many bad things came from standing terraces. All-seater stadia have been the catalyst for a new culture of safety at football grounds. There are new and different problems to be managed, but there's not enough evidence to go against Taylor on all-seater stadia.'

Sheila Spiers from the Football Supporters' Association (FSA) argues that, 'It is FSA policy that fans at every ground should have the choice to stand in small, modern terraces with no fences. There should be proper access and egress with computerised entry monitoring so that everybody is fully aware of how many people are in each section. There have been tremendous improvements in the way stewards and police observe crowds not just for behaviour but also for safety.'

It is clear that everyone accepts there can be no return to massive terraces like the Kippax, Holte End or the many Spion Kops. But terracing remains in the lower divisions and terracing can be safe. The 'Green Guide' to Safety at Sports Grounds recommends all-seating, but does allow for the possibility of safe terracing. To be read in conjunction with the Football Stadia Advisory Design Council's 1993 guidelines on 'Designing for Safe Standing at Football Stadia', the 'Green Guide' sets out the engineering and management means whereby safe modern terracing may be achieved.

There have been several technological innovations in support of safe terracing. The now well established computerised turnstile monitoring systems are configured so that the capacity of each area, each stand and the whole ground can be calculated and displayed by the computer counting the numbers of persons passing through the relevant turnstiles. Flow rates can be displayed as graphs, enabling management to predict whether the number of people still to gain entry to the ground can actually be admitted in the time left before kick-off. Decisions on delaying the start of the match can be taken accordingly. The systems can be programmed to trigger alarms when chosen percentage capacity of a particular area is reached, allowing stewards to be deployed in time to redirect people to alternative areas if necessary.

NNC's crowd monitoring system provides pressure sensors at strategic points such as on crush barriers or at entrances or exits. The sensors transmit measurements of pressures to a computer screen where control room staff can be alerted if danger levels are reached. The measurements can be used to build up a picture of crowd patterns to optimise safety at future games through the correct positioning of barriers, the design of access/egress routes and the deployment of stewards. NNC has, however, yet to find a stadium willing to invest in the new system.

Finally, while a return to safe terracing at all-seated grounds remains a theoretical – and for some desirable – possibility, it may actually prove impossible in practice. As de Quidt points out, 'There are engineering reasons why, in many cases, it would not be practicable to convert seated areas back to terracing. Firstly, the decks in many stands have not been designed to take crush barriers. Secondly, the concrete would have to be reconfigured with extra steps and modern stands will not have been designed to take the extra weight of concrete required. Finally, it would not be possible to increase the capacity because the entrances, exits and concourses could not accommodate additional people.'

6.4 – STANDING DEBATE

The original citation for this article is: Frosdick, S. (2003) 'Standing Debate', *Stadium & Arena Management*, June 2003, pp. 25-26.

There is an ongoing debate in the UK about the risks around spectators standing up in all-seated grounds. Steve Frosdick investigates.

To prevent any recurrence of a crush-related disaster, all-seated accommodation is a requirement for all matches played in the top two divisions of the football leagues. The same is true across Europe for all UEFA competitions. Many new all-seated stadia have been built and many others have been redeveloped. Where terraces do remain, these have been constructed or renovated to comply with modern safety requirements.

Figure 6.4.1 The Purple Stand at Walsall in the Midlands is a good example of a modern stand with both safe standing and seated accommodation.

Football grounds have become safer places, and removing the large terraced 'Kops' has been a major factor in this. John de Quidt is Chief Executive of the Football Licensing Authority (FLA) and often uses the example of Liverpool to remind people that, before the Kop was rebuilt, some 25 to 40 people were injured at every match. Nowadays only three to five people are treated throughout the whole ground, and then mostly for illness rather than injury.

But stadia have also become less exciting places. Once there were lively fan groups who were actively involved in creating the spectacle. Now there is a more docile audience who passively consume the event. All-seating is recognised as the major factor in this loss of atmosphere. Understandably therefore, there has been pressure from fans for the provision of standing accommodation. The call is not for a return to the larger old-style terraces, but for small standing areas, built to modern specifications, perhaps behind the goals.

The pro-standing lobby have campaigned hard and won much support. Approaching the 1997 election, even the Prime Minister, Tony Blair, was quoted as saying there was no reason to ignore the technological improvements which might allow for safe standing. There are lots of web sites on which the supporters' arguments are well set out. Try typing 'safe standing' into your favourite search engine and exploring some of the links.

As far as the fans are concerned, there is now a clear majority view. The Football Supporters Federation recently reported an official survey, commissioned by the Football Association, which showed that 62% of fans were in favour of having standing areas. So how have the authorities responded to this populist clamour for change?

A number of German football grounds have convertible seating systems, allowing terracing for domestic games and seating for UEFA matches. In a positive response to supporter pressure for an investigation, Kate Hoey (the then Minister for Sport and a personal advocate of possible change) asked the FLA to visit the Volksparkstadion in Hamburg in February 2001 to assess whether their 'Kombi' seating system could reasonably and practically be installed in England and Wales.

Safety and security at sports grounds 197

Figure 6.4.2 The German 'Kombi' seat allows rapid conversion from standing terrace to seating deck when required for UEFA matches. (Photos reproduced with the kind permission of the FLA).

The FLA report is available from www.flaweb.org.uk. It noted that the combination of 'Kombi' seats and removable barriers was an ingenious and well-engineered system which could, with certain modifications, comply with the required safety standards. However the report also argues a clear case that the costs and size of footprint required for a 'Kombi' equipped structure would make it impracticable anywhere other than a new-build stadium on a suitably large site. The FLA concluded that the 'Kombi' system did not affect the arguments for or against the retention of Government policy on all-seated grounds. Fair enough. At least the system had been properly looked at.

But the Government position has otherwise remained unchanged. Richard Caborn has replaced Kate Hoey as Minister for Sport and is not sympathetic to a change of policy. As recently as 28 November 2002, Hansard (the proceedings of the UK Parliament) recorded a Member of Parliament asking for 'a statement or preferably a debate on the growing feeling that it is time to consider introducing *safe standing areas at football grounds'*. As yet, there has been no response.

In parallel with this thwarted campaign for change, a phenomenon has emerged of fans persistently standing in seated areas throughout the game. Of course fans are often stood up before the match, at half-time and when waiting to exit at the end. And fans stand up at moments of excitement or tension. All of these occasions could pose public safety risks, particularly during egress, but nobody seems too concerned about them. But persistent standing <u>during</u> the match <u>is</u> seen as a problem by the authorities. Why?

In my view, the authorities perceive the risk because they see the fans are challenging them. This is part of a general spectator protest against enforced changes in the traditional culture of football supporting. In the case of standing, the logic runs something like this. The fans want to stand and have asked for a change of policy. The Government are unmoved. So the fans stand up anyway and tacitly challenge the authorities to make them sit down. Attempts to do so have met with vocal and sometimes physical opposition and some very unpleasant scenes have resulted. Local certifying authorities then add to the vicious spiral by threatening to cut ground capacities and thus the club's gate money. It's all been getting rather nasty.

So the authorities have been faced with a rather complex risk management problem. There are safety concerns about injuries from people falling over the back of seats in front and about gangways being blocked. There are concerns that the anti-authority stance of standing fans makes the crowd more difficult to manage. There are customer care issues around the viewing and comfort of those who do not wish to stand. And there are concerns that persistent standing encourages turnstile frauds – you can't tell how many people are actually in that area.

In an attempt to find a way through the maze, the FLA, in association with all the other relevant authorities, have published a guidance document on Standing in Seated Areas. The document is available from www.flaweb.org.uk/standing%20in%20seated%20areas.htm. It is a well-considered report, which recognises that there are no quick or easy solutions.

Above all, the report recognises that local circumstances are different; and that a process of local risk assessment should inform any action. Quite right, but risk assessment is a judgement call rather than an exact science. Different individuals and stakeholders perceive risk in different ways and so there is often more than one 'right' answer. And of course there are wide variations in circumstances. Some grounds have only a few people persistently standing at the back of one stand. Others have huge numbers involved. Some clubs are a particular problem away from home, in one case all standing in a steeply raked upper tier.

To get a feel for the complexity, let's look at three alternative ways of assessing the safety risk. We will use the following scale to assess the probabilities and consequences of injuries.

Probability	Very Unlikely	Unlikely	Possible	Probable	Very Probable
Consequences	Slight	Minor	Moderate	Serious	Severe

First, we could use hindsight to ask how often people have been injured as a result of persistent standing and how serious their injuries have been. According to my sources, the data is scant. Only a handful of incidents are known, although in one case a couple of fans sustained broken limbs. But thousands of fans persistently stand every week. So we might reasonably argue that the probability of injury was 'very unlikely' and that the consequences were no more than 'moderate'. We might thus assess the risk as 'low' and conclude that no particular management intervention was required.

Taking another view, we might argue that common sense tells us that, at moments of excitement, there is a good chance of people toppling over. If they did, this could clearly result in broken bones or even deaths as people were crushed through sheer weight of numbers. There may not have been a serious incident but it is complete folly to wait for a disaster to prove the point. On this analysis then, the probability of injury is at least 'possible' and the consequences potentially 'severe'. Thus we might assess the risk as 'high' and conclude that we should do everything reasonably practicable to stop it. If there was physical opposition to our efforts, if our stewards had to eject people and if people in the area were frightened by the consequent disorder, then so be it – that was the price to be paid.

A third view seeks to achieve a reasonable balance of risks. Here we might begin by acknowledging the evidence that injuries had occurred but conclude that, given the numbers involved, the probability was 'unlikely'. Considering the consequences, our judgement might be that the most realistic scenario was broken limbs and therefore 'moderate'. Thus we might assess the risk as 'medium' and needing some intervention. However, we might then consider the risks to the safety of our stewards and to the general atmosphere in the ground. Now we might conclude that staff injury, public disorder and fear were both 'probable' and potentially 'serious'. Thus our intervention would create more serious risks and so, on balance, we might choose to put up with the persistent standing.

Whilst researching this article, I initiated a discussion on the Football Safety Officers' Association website. The responses showed that these three hypothetical assessments can indeed be found in real-life. Reflecting the first view, one safety officer commented that he had about 100 fans who persistently stood in a home area with a capacity of about 1000. 'We initially gave out leaflets to the supporters in that stand', he said, 'and we then spent two days picking up the litter! We have not had any injuries for eight years and no one has complained.'

Another safety officer gave a detailed and persuasive analysis of the risks. He explained the comprehensive set of actions he had taken to address them and reported the opposition he had encountered. He concluded that, 'My name is mud but we are getting there. The situation is ten times better now than a year ago and the violent response to stewards has reduced considerably. We have lost some horrible supporters and retained some good ones. We have the full support of the Supporters Club and of many people in the relevant area of the ground. It's a two or three year job and hard work.'

Reflecting the third view, another safety officer reported that, 'At my club I have one home area that persistently stands. They are considered to be the ardent club supporters and are also the ones who go to away games. I initially went public on standing as they gave me problems when asking supporters to sit down in other areas. I went on local radio and ran features in the local paper and programme. I put out leaflets and also have signs displayed. However during my eight years at the club I have never had a single supporter injured by standing. I agree it is a customer care issue but all the supporters in that area know that standing occurs. The area if full contains 2500 supporters and I will not put my stewards in jeopardy by having confrontations between them and the supporters which could lead to violence.' Another safety officer commented that, 'If we have a large crowd who are determined to stand, I for one am not prepared to die in a ditch to make them sit down'.

This is clearly a thorny problem. Some safety officers have rightly decided to live with it, whilst others have rightly tackled it, notwithstanding that this requires sustained action over a long period of time. And it is entirely proper that such decisions are taken locally. Given the substantial variations in circumstances, it is in my view quite right that the authorities have, to date, allowed the clubs to take the lead in assessing and managing the risks. However there is growing evidence that the public authorities will soon choose to act to impose requirements for enforcement. Whether these will allow for local differences remains to be seen.

The public authorities would perhaps do better to reconsider the case for safe standing. Yes, there are significant engineering, technical and financial issues which in most cases preclude the conversion of existing seating areas. But we could retain safe terracing or permit its construction in new stadia.

The Purple Stand at Walsall provides an excellent example of the perversity of current requirements. At present, some 2700 fans enjoy standing safely in the lower tier. But because Walsall have been in the first division for three years, they are now required to convert the lower tier to all-seating with a reduced capacity of around 1900. And it's a fair guess that the fans will all stand up throughout the game. So the club will incur construction costs and loss of revenue and the fans will exchange a safe terrace for seats where they will stand in protest. The stewards and police will then be faced with public safety and order problems where none existed before. This is plain daft and surely not what Lord Justice Taylor intended.

Figure 6.4.3 Standing terraces still exist at lower league football grounds in the UK. Will they ever return to larger stadiums?

6.3 AND 6.4 – COMMENTARY BY JIM CHALMERS – PERSISTENT STANDING IN ALL-SEATED GROUNDS

Article 6.3, written by Frosdick in 1998, examines whether there is a case to be made for the re-introduction of standing terraces in UK Premier and Division One League stadia, which were required to become all-seated as a consequence of the 1989 Hillsborough stadium disaster. In Article 6.4, written in 2004, Frosdick continues this debate, this time with a focus on the relatively new phenomenon of football fans who persistently stand in all-seated areas. Since these articles were written, the Football Safety Officers' Association has defined 'persistent standing' as follows: 'supporters will be considered to be persistently standing should they continue to stand after being requested to sit by a steward, police officer or agent of the club'.

The issues raised in both articles will rumble on and on, but it is important to distinguish between the desire of some fans to see the return of standing terraces and the actions of those who persistently stand. The latter instance is seen by many as a protest intended to encourage regulators to relax the all-seated requirement; a protest which to date has failed to have any impact on either the government or the football authorities. The former issue is whether a case can be made out for the reintroduction of small standing terraces, since as Frosdick rightly says there has never been a call to reintroduce the 20,000 plus capacity standing terraces of yesteryear. Those fans who advocate the return of standing terraces would argue that, since safe standing terraces are permissible in the lower divisions of the Football League, in rugby grounds, at horse racing and at outdoor pop concerts, then it is hypocritical to suggest that such standing could not operate equally safely in the two upper divisions. My view is that this is a powerful argument and that such fans have a strong case.

Frosdick uses Bescot Stadium, the home of Walsall FC, as a case study to illustrate this point. As an inspector with the Football Licensing Authority I monitored spectator safety at this stadium between 1991, shortly after the new stadium was constructed, and 2003, when I retired. The lower tier of the Purple Stand shown in the article was constructed to fully comply with the 'Green Guide' to Safety at Sports Grounds and was later slightly altered to fully comply with the Football Licensing Authority criteria for safe standing terraces. To all intents and purposes therefore, this stand was constructed, maintained and tested to meet all the relevant guidance to ensure that it was a structurally safe standing terrace. During the twelve years I monitored safety at this stadium, not one spectator sustained injury as a direct consequence of the structure or the way it was managed.

From a historical perspective, therefore, this stand was a low risk environment. The probability of spectator injury was very low and should anything happen on the terrace the consequences would be minimal. So whilst Walsall FC played in the lower divisions, the regulatory bodies, including the Football Licensing Authority, were more than happy for fans to use this area as a standing terrace. However all of that changed when the club was promoted to Division One. Now the legislation required the club to become all-seated. But the fans had not changed, nor had their behaviour, nor indeed had the standing terrace become structurally unsafe overnight. I recall my lengthy discussions with the club chairman and with fans, and the difficulty I had convincing them of the need to change. To this day I do not think the chairman or the fans are convinced this was the right thing to do.

Thankfully the concerns expressed by Frosdick in his article have not materialised, with the fans quickly adjusting to their new all-seated environment. It remains a fact however, that if the Club were relegated back to Division Two, then there would be nothing to stop them taking out the seats, re-erecting the crush barriers and converting this area back to a standing terrace. I do not understand why in one division it is acceptable to stand, whilst in another division it is not acceptable. Since it would be technically feasible for some seated areas to be converted back to small standing terraces, I have come to the conclusion that safe standing terraces could be reintroduced at stadia. Whether they will be reintroduced or not is another matter.

I was one of the authors of the Football Licensing Authority 'Kombi-Seating Report' referred to in Article 6.4. This demonstrates one engineering solution used in Germany, but I am not convinced that this would work in the UK. For both comfort and safety reasons, I believe an area needs to be designed for either seated or standing use. Trying to combine both is not the solution, as a number of football fans and I saw at a recent demonstration of another type of engineering solution proposed by the Football Supporters Federation. On this occasion it was the fans who said it would never work.

Lord Justice Taylor's recommendations for all-seated stadia were the catalyst for all the other changes which have occurred in terms of crowd behaviour, management and control. We should not forget that the all-seated environment has made the game more accessible to women, children and persons with disabilities. In the early days of the Football Licensing Authority, I remember speaking to various supporters groups and forums. They vehemently opposed the all-seated requirement and accused me of introducing something which would be the death knell of football in the UK. They said that fans would protest by quitting the game in droves. However the prophets of doom were proved wrong. In the 1989 to 1990 season, the year of the Hillsborough disaster, total league attendances were 19,466,826. In 2003 to 2004, total league attendances were 29,197,510 – an increase of almost ten million. Most fans have mellowed towards all-seated stadia, with only a vociferous few still calling for a return to the old days.

Given this winning formula of increased attendances and improved comfort and safety, I think it unlikely that Government will ever relax the all-seated requirement for the top two divisions. That being the case the issue of fans persistently standing in seated areas will remain a topical concern.

There is evidence that the Football Licensing Authority publication 'Standing in Seated Areas at Football Grounds' is having some effect. This guidance document identifies the need for clubs, the Football Safety Officers' Association, the football authorities, local authorities, police and the Football Licensing Authority to all work together to address the problem.

To try and scope the problem, the Football Licensing Authority introduced a standard system for Clubs to report any persistent standing by home or visiting fans. This report is submitted to the local authority and forwarded to the Football Licensing Authority, which collates and analyses the data. Based on this evidence, the Football Licensing Authority concluded at the end of the 2003 to 2004 season that supporters at most clubs rarely, if ever, stood persistently and it was not an issue for Clubs in the lower two football leagues, many of which still retain standing terraces. In the Premier League and Division One, only five and three clubs respectively had fans who persistently stood at home, whilst six Premier League and two Division One clubs had fans who persistently stood away from home. There were also supporters who persistently stood at certain matches such as at local derby fixtures.

This monitoring process is ongoing but the real issue is how should it be tackled. Certainly at regional and national meetings of the Football Safety Officers' Association, the problem is a topic on every agenda. However, there is not one single solution which can cure it.

Clubs should produce a stadium safety management plan which will show how they intend to manage persistent standing, particularly their strategy for getting fans to sit down. The local authority responsible for the issue of the general safety certificate should then determine whether the Plan is adequate. If it is not, they should consider to what extent the stadium capacity should be reduced. There have been such capacity reductions but as members of my Association will testify, the application of such sanctions by local authorities has been very inconsistent. For example at one stadium the visiting supporters capacity was cut by over 2,000. In another ground, the capacity of a similar sized visiting supporters area was cut by less than 200. This inequality of enforcement may be prolonging the problem since there is no consistent message getting out to spectators.

The football authorities, including my Association, agree that education, publicity and persuasion is a better way forward than repressive or oppressive responses by either clubs or local authorities, since these could cause a backlash from the fans. The perfect example of this was seen at one game I attended in the Premier League, when out of 2,500 visiting fans approximately 100 persistently stood. At half time, the safety officer put out a public address message asking this group to sit down in the interests of safety and comfort. The safety officer then told them that, if they did not sit down, they would all be ejected – whereupon the entire capacity of 2,500 stood up and remained standing until the end of the game. A real example of people power in action, illustrating that force, mass ejections, cuts in capacities and threats are not the way forward in the long run.

I do not object to people's right in a democracy to seek the reintroduction of standing terraces. However, I do object to the anti-social behaviour which prejudices the safety and comfort of those who want to sit and watch the game. Persistent standing as protest undermines the case for safe standing. It will not influence those in government or regulatory bodies. It will, however, continue to be a complex problem which has to be managed. Clubs have to ensure the safety and comfort of everyone attending the event. The amount of time, energy and resources spent in managing the problem of persistent standing is totally disproportionate. It is time for the minority to solve the problem and just sit down.

6.5 – TEN YEARS OF SEATING

The original citation for this article is: Frosdick, S. (2004) 'Ten Years of Seating', *Stadium & Arena Management*, December 2004, p. 15.

Steve Frosdick marks the tenth anniversary of all-seating in British football grounds.

Before 1989, the British stadium industry was in decline. Grounds were sordid, crowds were dropping and football was blighted by hooligan violence. After the Hillsborough disaster, the Taylor Inquiry reports led to wide-ranging and radical change. Most notable change was the recommendation that clubs should have all-seated grounds. Taylor said that 'There is no panacea that will achieve total safety and cure all problems of behaviour and crowd control. But I am satisfied that seating does more to achieve these objectives than any other measure'.

Using powers set out in the Football Spectators Act 1989, the Secretary of State made all-seated stands a legal requirement for the top two Divisions, and a deadline was set for August 1994. The Football Licensing Authority (FLA) were given the task of overseeing the implementation, together with the responsibility for monitoring the work of the local authorities issuing ground safety certificates.

Some of the early terrace conversions were poor. At grounds such as Luton Town, cheap plastic seats were simply bolted onto the existing terraces. These were unsuitable and the were problems with breakages, poor sightlines and inadequate leg-room. At other grounds, for example Portsmouth, uncovered terraces were converted to uncovered seating. It is bad enough standing in the rain, but sitting on a wet seat getting soaked is even less comfortable. Elsewhere, for example at Manchester United, the rakes of the lower tiers meant that conversion to seating did not help the viewing standards for fans.

In general terms, however, the conversions and rebuilds have been positive. August 2004 saw the tenth anniversary of the all-seated deadline, and it is thus timely to celebrate the beneficial changes seen in British football grounds.

All-seating has meant greater comfort. A seat gives each fan their own personal space. This is particularly important for women, children and older men, who are less able to cope with the crowd pressures on a standing terrace. When building or converting stands, clubs have also taken the opportunity to improve comfort and generate new revenues by providing fans with decent facilities. Foul latrines and lukewarm pies have been replaced by modern sanitation and a wide range of food and beverage. Club seats and executive box facilities have also been widely provided.

All-seating has also resulted in improved viewing standards. The many new stands built since 1989 have had cantilevered or goalpost roofs to eliminate the restricted views from old roof props. Modern design has allowed for improved C values – the distance between your eyes and the top of the head of the fan in front – so that more people have a clear view of the action.

Above all, seating has brought about Lord Justice Taylor's vision of greater spectator safety. FLA Chief Executive John de Quidt often uses Simon Inglis' example of Anfield to remind people that, before the Kop was rebuilt, some 25 to 40 people were injured at every match. Today, only three to five people are treated throughout the whole Liverpool ground, and then mostly for illness rather than injury.

Certainly, the FLA has had a catalytic role in the change process. John de Quidt highlights how the FLA has taken the lead in driving all-seating through. The emphasis has been on providing better facilities as part of an integrated approach to safety, including allocation of responsibilities and customer care measures to help the fans feel safe and secure. There has been a particular focus on good management as well as good design.

All-seating has been adopted as a requirement for UEFA competitions and the oversight role of the FLA has attracted interest from elsewhere in Europe and the rest of the world. Both Brazil and the Czech Republic have introduced legislation learning from the English experience. Brazil are appointing a National Commissioner and the Czech Republic are giving an oversight role to their local authorities.

Of course these benefits have come at a price. Grounds are safer but the atmosphere is blander. The active participant on the terraces has been replaced by the more passive consumer in the stands. However this is a small price to pay for past disasters. There are arguments for the re-introduction of safe standing but there is no one who wants to see a return to the old-fashioned Kops.

Overall, all-seating has been an icon for the transformation of British football grounds into 21st century entertainment environments – more comfortable, with better viewing, but above all safer. We've come a long way in the last ten years.

6.5 – COMMENTARY BY JIM CHALMERS – THE TEN YEAR ANNIVERSARY OF ALL-SEATED STANDS

Frosdick reflects briefly on the tenth anniversary of all-seated stands in British football stadia in 2004. He outlines how the requirement came about and how the Football Licensing Authority has the statutory authority to enforce this requirement in the top two Divisions. I know Frosdick was commissioned to write only a brief summary, but the article does not do justice to what has been achieved, particularly by the Football Licensing Authority (FLA).

As one of the nine Football Licensing Authority Inspectors from 1991 to 2003, my colleagues and I had the day-to-day responsibility of ensuring clubs faced up to the challenge of change and complied with the statutory time limits to implement all-seating at their stadia. And what a challenge it was. 59 out of the 92 club grounds had been built between 1890 and 1910. Apart from the erection of high pitch perimeter fences in the 1960s and 70s they were virtually untouched at the time of the Hillsborough Disaster in 1989. A monumental task had to be addressed but the history of complacency in the world of football persisted even after the tragic deaths of 96 fans. Clubs said that the Hillsborough Disaster was a 'one-off' and would never happen at their ground. Despite the financial assistance available from the Football Trust most clubs pleaded poverty as a reason not to change. Only a few clubs accepted the need for change. Some club chairmen said the FLA would be disbanded before the 1994 deadline and that, even if they remained, they would never be able to force clubs to close their standing terraces.

So the world of football did not wholeheartedly embrace the principle of all-seating and some had to be dragged kicking and screaming in the new direction for our stadia. Overall therefore it was probably the most interesting, demanding, frustrating but enjoyable and rewarding time in the history of the FLA – playing a key role in seeing out the old and bringing in the new.

It was not however just about seeing the terraces replaced by seats. The FLA philosophy was about encouraging clubs to build to the maximum rather than the minimum dimensions, particularly with angles of rake, seat row depth and seat widths to provide the most comfortable view of the game. New roof designs were about getting rid of support pillars to remove obstructions to sightlines. The FLA role also extended into areas of customer care, such as the provision of well designed concourses with modern refreshment and toilet facilities.

Not every club embraced this philosophy of customer care and I recall one chairman saying to me, 'There is no point wasting good money on toilets, they will only smash them up'. Another said, 'All they want is a pie and Bovril and I don't need to spend thousands for that'. The terrace mentality still prevailed in some boardrooms as an excuse to do and spend as little as possible. This negative outlook has been proved wrong over the years as it has been shown that fans will respect and enjoy the highest standards of facilities. In terms of customer care, why should they expect any less? And there are superb examples of modern concourse designs with facilities to match. Examples in my area included the concourses in the new Holte End at Villa Park, Birmingham and the East Stand at Highfield Road, Coventry.

The FLA role, therefore, was not just about seeking compliance with all-seating. It was also about ensuring clubs did not replace a poor standing terrace with an equally poor seated stand. The FLA can reflect with some pride on their leadership, persuasion, coercion and threats at times, which have helped achieve the general high standards of spectator accommodation now seen in UK stadia.

In the ten years before Hillsborough only ten new stands were built in England and Wales. In the post Hillsborough era virtually every stadium has been touched by developments and improvements, with many clubs moving to a brand new stadium. Clubs which have relocated include: Chester City, Millwall, Huddersfield, Middlesborough, Derby County, Reading, Sunderland, Oxford United, Stoke City, Manchester City, Bolton Wanderers, Northampton Town, Leicester City, Hull City, Rushden and Diamonds, Wycombe Wanderers, Southampton, Walsall, Wigan Athletic, Darlington and Grimsby.

The process of development continues in 2005 with new stadia either planned or under construction for: Coventry City, Liverpool, Swansea City, Milton Keynes Dons, Arsenal, Brighton and Hove Albion and Shrewsbury Town. Of course the jewel in the England crown is the construction of the new Wembley Stadium. We should also not ignore the individual new stands still being planned for those clubs where relocation is not an option. It is estimated that in the post Hillsborough era well over 200 new stands have been built in England and Wales alone at a cost estimated to be in the region of £2 billion.

The process has not all been plain sailing. I would like to say that every new stand which has been constructed has been perfect, but sadly that is not the case. Stands have been built with the minimum seating dimensions, affecting comfort and viewing. There are those with bleak and limited concourses with basic toilet and refreshment facilities. The needs of people with disabilities have in some cases been an afterthought and even then ill considered with poor sightlines. Thankfully the poor new stands are in the minority. Overall the general standard of new spectator accommodation is high.

The transformation of UK stadia continues with hotels, conference and leisure facilities being incorporated into the designs. It could be said that phase one of the changes, immediately after Hillsborough, was to meet the statutory requirements for all-seating and, in the lower leagues, the provision of safe standing terraces. It could be argued that phase two of the developments reflect the continued popularity of the game, increased attendances, television revenue and the need for stadia to explore their full commercial potential in a 24/7 operation. The next decade will I am sure bring many more new and exciting developments in UK football grounds.

PART VII – EDUCATION

Introduction

Part VII comprises three articles dealing with the qualifications and competencies of sports grounds safety officers. Particular mention is made of the Certificate of Higher Education in Stadium and Arena Safety, pioneered in 1999 by the University of Portsmouth. Part VII concludes with a description of the UK involvement in sharing sports grounds safety experiences with students at the Université de Technologie Troyes, France.

7.1 – Certificate of Higher Education in Stadium and Arena Safety

Frosdick describes how in 1999 he introduced a Certificate of Higher Education in Stadium and Arena Safety whilst a lecturer at the University of Portsmouth. Comment is made about the Certificate from the perspective of a student who also explains why the Certificate is no longer available.

7.2 – Evidencing the Competence of the Safety Officer

Writing in 2001, Frosdick expands upon article 7.1, examining in detail the training of stadium safety officers and how vocational and academic qualifications can help demonstrate their competence. Comment is made about the lack of mandatory training, assessment and qualifications for safety officers and how and how these inadequacies can be remedied.

7.3 – Lecture Visits to the Université de Technologie Troyes

Frosdick explains his involvement in 2001 with the Chief Executive of the Football Licensing Authority in sharing UK stadium safety experiences with students attending the Advanced Specialist Diploma in Sports Engineering and Management at the Université de Technologie Troyes, France. This involvement has continued annually over five years and now includes practical input from the club safety officer at the Stade de l'Aube.

7.1 – CERTIFICATE OF HIGHER EDUCATION IN STADIUM AND ARENA SAFETY

The original citation for this article is: Frosdick, S. (1999), 'The New University of Portsmouth Certificate of Higher Education in Stadium and Arena Safety', *Football Management*, Winter 1999, p.2.

Steve Frosdick outlines the University of Portsmouth Certificate of Higher Education in Stadium and Arena Safety

Managing stadia and arenas so as to provide spectators with an environment in which they can watch events enjoyably and safely is a highly complex problem. It includes questions of strategic and operational management; architectural and engineering design; technological sophistication; health and safety; public order; customer care; and an understanding of how people behave in crowds. Above all, it involves a sound understanding of risk assessment and management. Maintaining public safety requires the specialist knowledge of competent professional persons – whether hands-on safety officers, the staff of regulatory bodies, or other interested parties. Health and safety competence may be adduced from knowledge, experience and training; and it is becoming clear that those involved will increasingly need to satisfy others regarding their professional education and qualifications.

The Certificate of Higher Education in Stadium and Arena Safety is a unique course, offered by the Institute of Criminal of Justice Studies at the University of Portsmouth. The programme was developed by Football Management contributor Steve Frosdick, following consultation with the stadium and arena safety industry. The course is open to everybody – whether current practitioners, regulators, those seeking a career in the industry or other interested parties. It aims to equip students with the ability to understand and apply relevant approaches, techniques and processes from both general management and the specialist field, and so learn to apply a more reflexive and critical approach to their role.

The course will be taught entirely by specially produced distance learning study materials delivered to students' home addresses. The first materials will include a study-skills package to help students through their studies and to prepare them for their assessments. Tutor support will be available both by telephone and e-mail.

Course assessment will be a combination of project work, assessed essays and an examination. These will be held very six months at local examination venues throughout the United Kingdom and will be essay type questions based on a topic list sent to candidates one month in advance of the examination. The examination timetable will give students the flexibility to complete the course in either 12 months (in exceptional cases), 18 or 24 months (for normal part-time study) or up to

36 months (for students who want to take their time). Accreditation of timely and relevant prior learning (APL) will be permitted in accordance with the University of Portsmouth's APL procedures.

The Certificate of Higher Education can be used as a stand-alone award, that is, as a qualification in its own right, which can be used to help demonstrate competence as a public safety professional; but it can also be used as part of a wider qualification. The course will lead to the award of 120 credits under the national CATS scheme (Credit Accumulation and Transfer). This normally counts as the equivalent of one third of an undergraduate degree; thus, should students wish to continue in their studies on an appropriate course at any UK University, they will be able to take the credits with them and so gain exemption from the first year.

The course will be structured in seven units;

- Study Skills (10 credits) – to cover learning to study, essay writing, examinations and writing-up projects;

- Personal and Professional Development Profile (30 credits) – including both a written self-evaluation and a work-based project, which will enable students to apply what they have learned in practice;

- Management Studies (20 credits) – including planning, problem solving, decision-making, organisation behaviour, personnel, training and development;

- Management of Risk (10 credits) – to cover the main physical, psychological, sociological and cultural theories of risk;

- Health and Safety (10 credits) – introducing the key elements of occupational health and safety at work;

- Sport and Safety Management (20 credits) – to study a selection of essays to increase understanding of the theoretical context in which students work;

- Guidance on Safety at Sports Grounds (20 credits) – including the UK Government's 'Green Guide' and other practical guidance on safety management, contingency planning and similar topics.

7.1 – COMMENTARY BY JIM CHALMERS – CERTIFICATE OF HIGHER EDUCATION IN STADIUM AND ARENA SAFETY

Writing in 1999, Frosdick explains the introduction of a new Certificate of Higher Education in Stadium and Arena Safety which he pioneered through the University of Portsmouth. The distance-learning course was principally targeted at those involved in the management of stadium and arena safety but was of equal value to those seeking to enter this career path. The course filled a gap in higher education provision on this topic and Frosdick hoped the course would be embraced by many safety officers.

When the certificate was introduced in 1999, I was then an inspector with the Football Licensing Authority. The course content presented an ideal opportunity for me to develop my knowledge and competencies. Thus at the age of fifty-nine, and with no formal education since leaving school at the age of fifteen, I embarked on the course of study outlined in the article. I would not pretend the course was easy, with the hardest part being learning how to research and write an academic essay. However the University's tutorial support helped me overcome the obstacles and adjust to a part-time student way of life. I was eventually awarded the Certificate in July 2001 after nearly eighteen months of fairly constant study and essay writing. Surprisingly the examination did not present the same difficulties as I had experienced in coming to terms with essay writing.

Was the course of benefit to me? The answer is an unqualified 'yes'. Study developed my education, knowledge and skills. It taught me how to conduct research and how to present the findings in a structured and cohesive fashion. It also helped in the sense that as an FLA inspector I was constantly encouraging safety officers and stewards to develop and demonstrate their professional skills and competencies by training and qualifications. Spending so much time preaching this principle it only seemed right to practice what I preached and to lead by example.

Sadly, whether due to apathy, cost or arrogance on the part of safety officers, the course was not well supported. Only nine certificates were awarded, all between 2000 and 2003, of which just five were obtained by persons who have held positions as safety officers.

When I embarked on the course of study, I had always hoped that this would lead to a Diploma and a Degree in Stadium and Arena Safety, but this was not to be. Due to the lack of interest in the Certificate qualification, the University of Portsmouth could not sustain it as a stand-alone award, never mind develop it for higher-level awards. The course subjects were therefore assimilated into the University of Portsmouth Diploma and BSc (Hons) in Risk and Security Management. I nevertheless continued with the course and will graduate with a degree in July 2005 at the age of sixty-five.

At Diploma and Degree levels students can still study the modules related to stadium and arena safety, but the course is in my opinion mainly aimed at police officers and those employed in the security industry. However this should not dissuade anyone involved in stadium and arena safety since the course of study can only improve their professional knowledge and skills.

I still wish there were specific higher educational qualifications for venue safety managers or those wishing to explore this career path, particularly at certificate level, which would have been well within the capabilities of any safety officer at whatever age or educational background. Sadly what Frosdick and a few students pioneered is now subsumed in a different academic strand. As a current sports ground safety manager, I can say that we have only ourselves to blame for this loss.

7.2 – COMPLETELY SAFE

The original citation for this article is: Frosdick, S. (2001) 'Completely Safe', *Panstadia International Quarterly Report*, Volume 7 Number 3, January 2001, pp. 70 and 72.

How can the competence of the public assembly facility safety officer be evidenced? Steve Frosdick outlines the debate.

In the July 2000 edition of 'Panstadia International', Pat Carr reviewed the provision of training in security, stewarding and crowd management procedures for stadia events in the UK. He argued that training was particularly vital for the safety officer and questioned whether being a retired police officer was sufficient qualification for this demanding role. He wrote that 'the sooner a recognised degree level course specifically for stadia safety officers is developed, and then made mandatory prior to taking on the role, the quicker we will progress to a higher nationwide/stadia-wide standard'. As Course Leader for the University of Portsmouth's Certificate of Higher Education in Stadium and Arena Safety, Pat Carr's article was of course music to my ears!

Managing stadia and arenas so as to provide spectators with an environment in which they can watch events enjoyably and safely, is a highly complex problem. It includes questions of strategic and operational management, architectural and engineering design, technological sophistication, health and safety, public order, customer care and an understanding of how people behave in crowds. Above all it involves a sound understanding of risk assessment and management. Given this complexity, Pat Carr is quite right to question whether a previous police career will have provided an aspirant safety officer with the breadth of knowledge and skills required.

Maintaining public safety requires the specialist knowledge of competent professional persons, whether 'hands-on' safety officers, the staff of regulatory bodies, or other interested parties. Within the UK, the debate about evidencing the competence of the safety officer is gathering momentum and it is becoming clear that those involved will increasingly need to satisfy others regarding their professional education and qualifications. Competence may be adduced from knowledge, skills and experience. Knowledge is acquired through education and experience. Skills are acquired through training and experience. Experience alone is no longer enough.

Within the UK, we can trace three main approaches to evidencing such competence: through continuing professional development, through vocational qualifications and through academic study. Continuing professional development may be evidenced through attendance at relevant seminars and conferences, which are offered by a range of institutions. For example, the Home Office Emergency

Planning College run a three day seminar on 'Safety at Sport and Entertainment Venues'. And the Football Safety Officers' Association hold two formal national meetings each year. More internationally, the European Stadium Managers Association (ESMA) convenes an annual conference, which usually includes one or more papers on safety issues, and the International Association of Assembly Managers (IAAM) in the United States host an International Crowd Management Conference in November each year.

But *continuing* professional development implies some building on an initial foundation of evidenced competence, either through vocational or academic qualifications. Safety officers who wish to follow the vocational path may choose to compile a portfolio of evidence to meet the requirements of the Level Four National Vocational Qualification (NVQ) in Spectator Control. On the more academic path, the Certificate of Higher Education in Stadium and Arena Safety is a unique course, offered by the Institute of Criminal Justice Studies at the University of Portsmouth. The programme was developed in consultation with the stadium and arena safety industry and is open to everybody, whether current practitioners, regulators, those seeking a career in the industry or other interested parties. It aims to equip students with the ability to understand and apply relevant approaches, techniques and processes from both general management and the specialist field, and so learn to apply a more reflective and critical approach to their role.

Its design and delivery make the Portsmouth programme unique. Looking world-wide, more generic stadium management programmes, perhaps aimed at senior executives, are available through several sources. The IAAM offer a Public Assembly Facility Management School which includes crowd management but covers a whole range of other stadium management topics. Students attend for one week each year over a two year period. A similar programme is offered in Australia by the Asia/Pacific based Venue Managers Association (VMA). The World Council for Venue Management (WCVM), which is the umbrella organisation for associations including the IAAM, VMA and ESMA, also offers Masters and even Doctoral qualifications in general public assembly facilities management, but the words 'safety' and 'security' do not feature specifically in these general management qualifications. Elsewhere in UK academe, a search of the Universities and Colleges Admissions Service website (www.ucas.ac.uk) reveals only one relevant hit. Liverpool Hope University College are planning to offer a full time combined honours programme in Health and Safety Management with Sports Studies from 2001, but it is not clear whether this will include safety at sports grounds.

There is, however, no other known undergraduate course which is specifically aimed at stadium and arena safety management and which does not require attendance at the university. A key feature is that the Portsmouth course is taught entirely by specially produced study materials delivered to students' home addresses. It is thus a distance learning course.

The course is structured in seven units: Study Skills; Personal and Professional Development Profile; Management Studies; Management of Risk; Health and Safety; Sport and Safety Management; and Guidance on Safety at Sports Grounds.

Tutor support is available by telephone, e-mail and at study schools, which are held twice a year. Course assessment is a combination of project work, assessed essays and examinations. The timetable gives students the flexibility to complete the course in either 12 months (in exceptional cases), 18 or 24 months (for normal part-time study) or up to 36 months (for students who want to take their time). Accreditation of timely and relevant prior learning (APL) is permitted. For example, students who hold the Level Four NVQ in Spectator Control are exempted from 40 credits.

The Certificate of Higher Education can be used as a 'stand alone' award, i.e. as a qualification in its own right, which can be used to help demonstrate competence as a public safety professional. But it can also be used as part of a wider qualification. The course leads to the award of 120 credits under the national CATS scheme (Credit Accumulation and Transfer). This normally counts as the equivalent of one third of an undergraduate degree, thus, should students wish to continue in their studies on an appropriate course at any UK University, they are able to take the credits with them and so gain exemption from the first year.

Pat Carr should be pleased to know that there is a recognised university course specifically aimed at stadia safety officers. There is of course also the Level Four NVQ. The question of whether either of these should be mandatory will no doubt be addressed in the ongoing debate about evidencing competence in stadium and arena safety management.

7.2 – COMMENTARY BY JIM CHALMERS – EVIDENCING THE COMPETENCE OF THE SAFETY OFFICER

The article written by Frosdick in 2001 followed an article written by Carr in the July 2000 edition of 'Panstadia International' magazine. Carr called for a recognised degree level course for stadium safety officers. He felt this would lead to higher standards of competency and argued that such a qualification should be mandatory prior to anyone taking on this role. In support of this argument, Frosdick refers to the Certificate of Higher Education in Stadium and Arena Safety from the University of Portsmouth. Much of his response to Carr relates to article 7.1, which has already been commented upon.

Frosdick does however extend the debate on safety officers being able to demonstrate their competencies through continual professional development, vocational qualifications and academic study. I agree with him when he describes how competency can be adduced from knowledge, skills and experience and how this can be demonstrated. I particularly agree with Carr where he argues that being a retired police officer should not in itself be regarded as sufficient qualification to be appointed as a safety officer. The role is far more complex than simple crowd control. There are numerous examples in all sports of where a retired police officer has been appointed as safety officer without the necessary breadth of competency. This is unfair on the post holder and can compromise the safety of supporters.

The police commander role is difficult enough in itself. In recommendation 52 of his report into Hillsborough, Lord Justice Taylor said, 'Consideration should be given to the provision of a specific training course for senior officers presently acting as police commanders and those in line to do so. Such a course should include training in the basic strategy of policing football matches'. In response, the police developed the 'Police Major Sporting Events Course', which is still arranged at various regional police training centres in England and Wales. Over the years, in the absence of a specific training course for them, many safety officers attended this course. This was not ideal, since the course was principally intended for operational police officers.

In Part II (article 2.1), the development of the Football Safety Officers' Association and safety officer training was commented upon in detail. I do not necessarily agree with Carr that an appropriate degree qualification should be mandatory prior to the appointment of a safety officer. If this was imposed it would mean the vast majority of our current safety officers would be considered unsuitable for the post and there would be no pool of suitable applicants to replace them. In reality, therefore, such a proposal would effectively shut our sports grounds down.

However the Football Safety Officers' Association six day 'Event and Matchday Safety Management Course', accredited by the University of Portsmouth, should be regarded as the essential minimum qualification for anyone who is currently a safety officer or deputy or is seeking to be appointed to such a post. The Taylor principle of training for the 'police commander' was accepted, so surely the same principle should apply to safety officers, who have the title of 'match commander' on the day of the event.

Safety officers can show continuing professional development by belonging to an appropriate safety officers association and by attending their meetings and conferences. They can attend various health and safety courses. They can attend seminars, which in the world of football have been arranged by the Football League and the Football Licensing Authority. However in terms of vocational and academic qualifications there is a marked reluctance by stadium safety officers to go down either path.

In Article 2.1 reference is made to the small number of safety officers (about thirty) in all sports who have attained the Level 4 NVQ in Spectator Control. Even fewer have gone down the academic path with only five safety officers attaining the Certificate of Higher Education in Stadium and Arena Safety referred to in article 7.1. Only two people have obtained a Level 4 NVQ in Spectator Control, the Certificate of Higher Education and the BSc degree in Risk and Security Management. Some safety officers possess other types of vocational and academic qualifications but they are in the minority.

The fourth edition of the 'Green Guide' to Safety at Sports Grounds describes how a safety officer will be regarded as competent when he or she has sufficient training, experience and knowledge to be able to carry out the functions set out in their job description. However no one has ever defined what sufficient training means nor have any training or qualification requirements ever been imposed on a safety officer through the General Safety Certificate issued to a stadium by a local authority. There is such a requirement placed on stewards to be trained and assessed to a nationally recognised standard – currently either the Level 2 NVQ or the Football Stewarding Qualification. This is enforced through the General Safety Certificate. It is somewhat perverse that there are strict criteria for the selection, training and assessment of stewards, but that no such criteria exist for the safety officer.

It is therefore my view that sports ground regulatory bodies should make it mandatory for all safety officers and their deputies to attend the 'Event and Matchday Safety Management Course' and then attain a Level 4 NVQ in Spectator Control. This requirement should be enforced through the General Safety Certificate and would enable some evidencing of competency as envisaged by Carr and Frosdick.

Perhaps the last word should be left to an ex police officer who, without any previous safety management training or experience, was appointed safety officer with a major rugby league club. He contacted me to say, 'I just wanted to thank you for advising me to attend the event and matchday course. It has made me realise that I knew nothing about the roles, responsibilities, liability and accountability which are what being a safety officer is all about. I would say that every safety officer should be made to attend this course since without it their competency for the post must be in question'.

7.3 – SAFETY TECHNIQUES SHARED

The original citation for this article is: Frosdick, S. (2001) 'Safety Techniques Shared', *Stadium & Arena Management*, June 2001, p. 4.

Having prepared for and built on the success of the 1998 World Cup, France is one of several European countries to have made significant advances in the management of stadium and arena sports. This is true both of governing and regulatory bodies and in the venues themselves. More sophisticated management demands better educated managers and the Université de Technologie Troyes – one of only six leading edge technology universities in France – has designed and developed a specialist education course to meet this need.

The Diplôme d'Études Supérieures Spécialisées Ingénerie et Management du Sport (Advanced Specialist Diploma in Sports Engineering and Management) is delivered as a post graduate or post professional experience programme and is structured in three stages. First is a study module covering general management in the context of sport. Second is a choice of engineering study modules. The concept of engineering used is very broad, including underpinning knowledge in subjects such as crowd psychology, best value for money and project management as well as more traditional subjects such as construction, maintenance and materials. The final stage of the course is a five month professional experience placement either in France or abroad.

Safety and security is also included and the University decided to invite John de Quidt (Football Licensing Authority) and Steve Frosdick (University of Portsmouth) to visit and share the lessons learned in Britain over the last ten years. The prospect of a trip to the Champagne region to talk about their favourite subject had a certain appeal so a programme for the visit was devised. A morning of lectures was to be followed by an afternoon 'hands on' visit to the Stade de l'Aube – home of ESTAC, the Troyes Premier Division side – and an evening conference open to the public.

John de Quidt lectured the Diploma students on the need for an integrated approach to spectator safety. The French word 'sécurité' covers both safety and order and John highlighted the need to distinguish between these two elements, maintaining a balance between them through the use of safety management systems, safety personnel and information and communications technologies. Other key points included crowd behaviour, stadium design and the stadium environment. Steve Frosdick then lectured on the history of disasters and disorder in British sports grounds. He went on to review the changes since the Hillsborough disaster in 1989, particularly the complex regulatory framework for safety, the lower profile of the public police and the emergence of more professional stewarding and safety management. He concluded by setting safety and security in a broader risk management context. The lectures were followed by a lively question and answer session.

The afternoon visit provided an enlightening opportunity for the students to tour the stadium in small groups, examining the venue to assess the risks to safety and security, to spectator enjoyment, to commercial profitability and to the local community. The students certainly learned a lot from applying the theoretical concepts from the morning lectures and the television station France 3 were present to record the study visit and interview the British visitors.

The well-attended evening conference was hosted at the university by Dr Patrick Laclémence, police commander with the Compagnie Républicaine de Sécurité (CRS) and an external examiner on the Diploma course. John de Quidt and Steve Frosdick summarised their morning presentations and Patrick Laclémence spoke about violence in the stadium, which he conceptualised as both a place of passion and a terrain of tension. An open forum concluded the conference.

The British visit was considered a great success from all sides. As well as forging new international relationships and giving the visitors the chance to practice their French language skills, it also encouraged both the students and the wider public to consider broader yet more integrated concepts of safety and risk management in the stadium. There was an appreciation that safety and security had come a long way, even since the World Cup, but there was more that could be done. There were also great opportunities to tap the commercial potential of French venues and to improve the spectator experience.

7.3 – COMMENTARY BY JIM CHALMERS – LECTURE VISITS TO THE UNIVERSITÉ DE TECHNOLOGIE TROYES

The article written by Frosdick in 2001 provides the background to how he and John de Quidt (Chief Executive of the Football Licensing Authority) came to be involved in teaching students on the Advanced Specialist Diploma in Sports Engineering and Management at the Université de Technologie Troyes, France. Their annual involvement with the University has continued to this day. I would suggest that the French students are very fortunate since the combination of Frosdick's and de Quidt's technical and management knowledge is quite exceptional.

With no disrespect to either Frosdick or de Quidt, they are an academic and a civil servant respectively. The fact that neither has ever taken control of an event day safety management operation could be seen as a weakness in presenting the students with the complete picture. For this reason, de Quidt and Frosdick welcomed the appointment by ESTAC in 2002 of a full-time club safety officer. This means that a practical expert is now also available to inform the students during the 'hands on' visits to the Stade de l'Aube.

APPENDIX 1 - MANAGEMENT RESOURCES

Books

Frosdick. S. and Marsh, P. (2005) *Football Hooliganism*. Cullompton: Willan Publishing (ISBN 1-84392-129-4).

Frosdick, S. and Walley, L. (eds.) (1999) *Sport and Safety Management*, Oxford: Butterworth-Heinemann (ISBN 0-7506-4351-X).

Official Guidance and Reports

Accessible Stadia, London: The Football Stadia Improvement Fund and the Football Licensing Authority, 2003 (ISBN 0-9546293-0-2).

Control Rooms, London: Football Stadia Improvement Fund and the Football Licensing Authority, 2005 (ISBN 0-9546293-1-0).

Department of National Heritage and Scottish Office, *Guide to Safety at Sports Grounds*. London: Stationery Office, 1997 (ISBN 0-11-300095-2).

Health and Safety Executive, *Managing Crowds Safely* (Second Edition), Sudbury: HSE Books, 2000 (ISBN 0-7176-1834-X).

Home Office (1990), *The Hillsborough Stadium Disaster 15 April 1989 - Inquiry by the Rt Hon Lord Justice Taylor - Final Report*. London: HMSO (ISBN 0-10-109622-4).

Training Package for Stewarding at Football Grounds (Further Revised Second Edition), Preston: The Football League, The Football Association, The FA Premier League and The Football Safety Officers Association, 2005.

The Football Licensing Authority web site at *http://www.flaweb.org.uk* includes the following publications:
- Guidance notes for Drawing up a Statement of Safety Policy for Spectators at Football grounds;
- Safety Certification;
- Contingency Planning;
- Exercise Planning;
- Briefing/Debriefing;
- Standing in Seated Areas at Football Grounds; Kombi Seating report.

INDEX

Symbols

11 September 2001 12, 74

A

abuse 70, 119, 126, 128, 129, 136, 138, 152, 153, 154, 155, 156, 160, 174
abusive chanting 31
academic 7, 8, 11, 18, 74, 78, 79, 82, 97, 105, 131, 158, 177, 210, 213, 214, 215, 216, 218, 219, 223
access control 96
access control card 91
accountability 4, 12, 14, 15, 16, 17, 18, 19, 20, 21, 36, 99, 105, 167, 173, 220
action plans 15
active/passive 186
active foresight 171
advertisers 25
advisory group 49
Advisory Group Against Racism and Intimidation 51
aggressive dissent 135
aggro 78
air intake ducts 103
ALARP 98
alcohol 5, 12, 69, 74, 78, 84, 85, 86, 87, 88, 89, 90, 124, 127
alcohol controls 87
alcohol sales 85, 88
alienation 119
all-seated stadia 5, 13, 177
all-ticket 29
allegiances 130
ambulance service 70
anarchic 187
anthrax 99, 101, 103
anti-racist measures 151
anti-social behaviour 77
architectural form 184
Argentina 120
Argentinean League 76
Arizona State University stadium 102
arrests 133
arrest statistics 82
Arsenal 32
assessment 158, 210
Association of Chief Police Officers 55
Association of County Councils 142
Association of Premier and Football League Referees and Linesmen 134
atmosphere 5, 13, 179
atmosphere areas 180
attacks on officials or players 26
attrition rate 77
away entrance 127
away fans 66, 124
away team coach 171

B

banner 23
banning orders 121
barcode 94
beer 85
behaviour of players 134
biological attack 101
blacklist 93
blame and claim 19
Board of Directors 16
bomb dogs 102
bomb scares 31
Bradford 15, 31, 46, 55, 59, 99, 114
bravado 129
breach of the peace 138
briefing 145
British Security Industry Association 140
building regulations 45
Building Research Establishment 65
building site 44
bunting 182
buses 124

C

caged 119
capacity 31
cathedrals of sport 185
CBRN 106
CCTV 4, 9, 12, 30, 66, 121

Centre for Public Services Management and Research 15
Centre for Risk Management 143
Centre for Football Research 134
Certificate in Event and Matchday Stewarding 160
Certificate of Higher Education in Stadium and Arena Safety 6, 9, 13, 210
Certifying Authority 49
Chairman 16
chanting 132, 152
cheerleaders 182
civil damages 16
climate of risk 99
Club Card 93
coaches 130
coin-throwing 126
commercial manager 23
commercial perspective 167
commercial viability 22
community 37, 45
complacency 172
complaint 154
concourse 28
conflict management 140
constructing 4
construction vehicles 45
contingency plans 25, 39, 108, 145
control room 50, 68
copyright legislation 33
cordon 122, 126
corporate killing 16
cost savings 167
COTASS 5, 12
counselling services 173
court 16
Croke Park 31
crowd 4
crowding 34
crowd reactions 139
crowd safety guidance 16
crowd segregation 5
CS spray 4, 12, 61
customer care 211

D

danger 4
database 94
Data Protection Acts 59
debriefing 154
demolition 42
derby matches 122
desegregation 108, 130, 131, 132, 133
design 4
detection 56
deterrent 121
disability discrimination 151
disabled 34, 98
disorder 130, 134
disruption 4
dissention 136
distance learning 211
documented risk assessments 173
drummer 181
drunkenness 86
drunken fans 87
dry moats 111

E

education 13, 112, 205, 210, 211, 215, 221
ejection 68, 80
elastic fence 111
electrified 119
emergency aid 151
emergency services 37
engineering drawings 36
engineering structures 130
Eric Cantona 26
European Championships 59
European Community 17
European Stadium Managers Association 216
Euro 2000 162
Euro 96 117
evacuation 31, 148, 149
Event and Matchday Safety Management Course 53, 218
evidence 17, 20, 22, 47, 64, 78, 81, 116, 134, 136, 145, 155, 160, 167, 174, 179, 189, 192, 193, 199, 204, 216

Ewood Park stadium 5, 179
executive boxes 29
executive lounge 66
explosions 31
exterior concourse 123
external alley 122
external disruption perspective 26

F

face painting 182
facial recognition software 103
fault-lines 80
female proportion 190
fences 12, 25, 26, 48, 99, 108, 110, 111, 114, 115, 116, 117, 119, 120, 121, 122, 130, 132, 163, 168, 169, 170, 193, 208
FIFA 85, 110, 111, 114, 117, 164
finance 4
fire 23, 30, 44, 46, 49, 55, 68, 71, 97, 99, 105, 114, 138, 144, 149, 151, 164, 167, 172, 180, 191
fireworks 26
fire risk 180
fire safety 30, 151
Fire Safety and Safety of Places of Sport Act 143
fire safety patrols 144
flags 24, 180
flares 34
floodlighting 31, 185
football hooliganism 80
Football Licensing Authority 9, 10, 12, 19, 37
Football Research Unit 63
Football Safety Advisory Group 38, 70
Football Safety Management 4
Football Safety Officers Association 8, 10, 20
Football Stadia Advisory Design Council 142
Football Supporters Association 63
Football Trust 147
forgery 77
Fratton Park 122

Friends Provident St Mary's Stadium 122
FSQ 159
fun 4

G

gates 26, 31, 68, 71, 99, 102, 110, 111, 117, 119, 129, 171, 172, 173
gate stewards 111
Greater Manchester Police 142
Green Guide 16
ground regulations 112
ground safety certificate 174
Guatemala 120
Guide to Safety at Sports Grounds 11
Guide to the Appointment, Training and Duties of Football League Club Stewards 147

H

hard liquor 85
hazards 14, 25
hazards register 36
hazards to sightlines 25
Hazard and Operability Study 12, 47
Health and Safety Commission 142
Health and Safety Executive 142
Health and Safety procedures 173
heating and ventilation systems 103
heightened awareness 104
Heysel 15
higher education 213
high visibility reflective jacket 171
Hillsborough 12, 15
holistic approach 24
Home Office Emergency Planning College 215
Home Office Police Scientific Development Branch 55, 59
hooliganism 48
hotel 42
hotel bedrooms 42
Human Rights Act 132

I

identification 56
inappropriate language 152
incident cards 154
indiscipline 134
inflammatory gestures 134
injuries 126
Institution of Structural Engineers 142
insurance 17
integrated ticket system 5
internal concourse 124
International Association of Assembly Managers 216
International Crowd Management Conference 216
International Institute of Risk and Safety Management 8

J

job-related education 6

K

Kick It Out 151

L

League Managers Association 136
legislation 17, 19, 20, 33, 37, 53, 74, 80, 89, 98, 99, 112, 120, 121, 141, 149, 164, 169, 173, 175, 176, 203, 207
liability 14, 15, 16, 17, 22, 35, 99, 109, 134, 164, 167, 173, 220
licensing laws 70
life safety systems 50
light pollution 32
liquor licence 86
Local Authority Safety Group 70
local residents 31, 45
Lord Justice Taylor 48

M

maintenance 30, 37, 41, 56, 98, 114, 221
Major League Baseball 101
management responsibility 53
managing risk 15
manslaughter 16
mascot 182
match commander 219
match officials 135
Mauritius 77
media 172
medical 149
medical centre 50
meeting and greeting 171
merchandising 29
metal detectors 102
Metropolitan Police Service 142
Middlesbrough Football Club 61
military operations 147
misdemeanours 77
missile throwing 48
moats 120
Monica Seles 26
monitoring 56
multi-agency 69
musical instruments 181
musicians 181

N

National Basketball Association 101
National Criminal Intelligence Service 50
National Football League 101
national ticketing 92
negligence 16
netting 122, 130
Newcastle United versus Sunderland 23
next of kin 172
noise pollution 32
non-match days 42
NVQ 148
NVQ in Spectator Care and Control 54, 161

O

operational conflicts 23
Oregon Arena Corporation 101
organised/unorganised 186
organ music 188

P

Panstadia 15
parking 23, 33, 34, 38, 69, 70, 171, 175, 176
park and ride 124
participate 180
performers 26
perimeter advertising 25
perimeter cordons 111
perimeter fence 110
perimeter hoardings 26
perimeter obstructions 26
persistent standing 6, 177
PFA 134
physical barriers 110
pillars 41
pirate merchandise 33
pitch 5
pitch incursion 77
pitch invasion 48, 109
pitch perimeter security 120
pitch side incidents 138
plastic glasses 88
player behaviour 12, 108
player misconduct 5, 138
police 4
police-free 170
police commander 8, 119, 218
police control room 169
police counter terrorist security advisor 106
police horses 70
Police Major Sporting Events Course 218
Police Scientific Development Branch 55, 59
policing 15
Policing Football Hooliganism 131
policing style 142
Portsmouth 122
post traumatic stress disorder 173
pre-planned celebrations 134
prejudice 152
print at home 96
prison 136
private contractors 143

private security industry 143
Private Security Industry Act 146
professional development 215
professional footballers 136
professional stewarding 150
publicity 122
publicity campaigns 103
public address system 70
Public Assembly Facilities 4, 14, 28
public order 48, 170
public transport 33
Purple Guide 16
pyrotechnics 26

Q

qualifications 210
quality of life 45
queue 34, 93, 129

R

race circuit 118
racial abuse 136
racial awareness 140
racism 5, 151
racist chanting 48
radio station 187
recognition 56
record keeping 144, 172
redevelopment 4, 43
Red Book 154, 158
referee 118
regulation 11, 65, 74, 89, 164, 165
report 154
residential streets 38
responsibility 4
rhythm 187
risk 4
risk assessment 4, 15, 36
risk management 36
risk management plans 35
rivalry 131
robberies 31
Royal Institute of British Architects 185
Rugby Union Premiership 59

S

safety and commercialism 24
Safety at Sports Grounds Act 143
safety certification 42, 53, 157
safety culture 67
safety features 149
Safety in Sports and Leisure Programme 18
safety management 149
safety managers 15
safety techniques 6
scapegoating 16
scoreboards 112
screens 112
searches 102
searching spectators 69
search and eject 167
season ticket 96, 179
seating 5, 13, 40, 44, 65, 98, 111, 117, 124, 177, 178, 189, 190, 192, 193, 196, 197, 200, 206, 207, 208, 209
seat number 156
security 5, 12, 102
Security Industry Authority 146
segregated zone 66
segregation 5, 12, 108, 118
segregation fence 125
Serie A 179
serious injury 110
sightline 29
signage 34, 70, 87, 88, 103, 122
signature tune 187
singing 124, 156, 180, 181, 182, 187, 189, 191
sin bins 139
site supervision 144
smartcard 96
soccer hooligans 78
Sound and Communication Industries Federation 142
Southampton 122
SpecIntell 54
spectator education 112
spectator experience 222
spectator perspective 25
spectator safety policy 19
spectator violence 4, 74, 76
sponsors 25
Sport, Space and the City 185
Sport and Safety Management 11
spotters 50
stabbing 26
Stadium and Arena Management 10, 28
stadium control room 169
stadium design 4, 65
stadium environment 221
stadium layout 149
Stadium Monitoring Group 38, 70
stadium structures 170
Staffordshire County Council Emergency Planning Unit 106
Staffordshire University 15
staff assessment 144
standards 51, 57, 98, 142, 143, 144, 147, 149, 178, 197, 206, 207, 208, 209, 218
standing accommodation 40
standing terraces 13, 177
standing up 5
statement of intent 19
stewarding 5, 9, 12, 17, 18, 21, 49, 51, 53, 59, 62, 69, 70, 80, 87, 90, 110, 111, 120, 121, 130, 131, 132, 140, 142, 143, 144, 145, 146, 147, 150, 157, 158, 160, 161, 162, 164, 166, 167, 168, 169, 170, 173, 180, 191, 215, 221
Stewarding and Safety Management at Football Grounds 58, 145
stewarding guidelines 5, 140
stewarding services 143
stewards 15
stewards training 5, 140
strategic risk 12
strategic risk management 8
streakers 115
street theatre 187
students 211
Super Bowl 102
supporters' representatives 39
switchboard 172

T

Talledega Speedway 102
Tartan Army 81
Taylor Reports 50
team benches 26
television exposure 23
terracing 43, 189
terrorist 105
them and us 119
threatening behaviour 132
throwing missiles 31
ticketing 12, 29, 41, 74, 87, 91, 93, 94, 95, 96, 112, 162
ticket office 69
ticket touting 77
tifosi 79
trainers 158
training aid 150
training materials 148
Training Package for Stewarding at Football Grounds 8, 21, 151
travel restrictions 80
tribal posturing 132
turnstiles 34, 44, 55, 66, 86, 87, 94, 96, 124, 128, 171, 193

U

UEFA Cup Final 76
UEFA Stadia Committee 117
ultras 79
unconventional threat 107
understanding crowds 4
Université de Technologie Troyes 6, 13, 210
University of Portsmouth 210
unlawful detention 132

V

verbal abuse 126
viewing distance 29
viewing location 29
Villa Park 59
violent behaviour 4
vocational 216

W

West Ham United 4
wheelchair users 159
whitelist 93
Winter Olympics 76
witness statement 155
World Series 102
written record 154

Z

zones 4, 24

Printed in the United Kingdom
by Lightning Source UK Ltd.
106076UKS00001BA/52-117